Luminos is the Open Access monograph publishing program from UC Press. Luminos provides a framework for preserving and reinvigorating monograph publishing for the future and increases the reach and visibility of important scholarly work. Titles published in the UC Press Luminos model are published with the same high standards for selection, peer review, production, and marketing as those in our traditional program. www.luminosoa.org

The Almond Paradox

The Almond Paradox

Cracking Open the Politics of What Plants Need

Emily Reisman

UNIVERSITY OF CALIFORNIA PRESS

University of California Press
Oakland, California

Suggested citation: Reisman, E. *The Almond Paradox: Cracking Open the Politics of What Plants Need*. Oakland: University of California Press, 2026. DOI: https://doi.org/10.1525/luminos.252

Cataloging-in-Publication data is on file at the Library of Congress.

ISBN 978-0-520-42306-0 (cloth)
ISBN 978-0-520-41383-2 (pbk.)
ISBN 978-0-520-41384-9 (ebook)

GPSR Authorized Representative: Easy Access System Europe,
Mustamäe tee 50, 10621 Tallinn, Estonia, gpsr.requests@easproject.com

35   34   33   32   31   30   29   28   27   26
10   9   8   7   6   5   4   3   2   1

publication supported by a grant from
The Community Foundation for Greater New Haven
as part of the *Urban Haven Project*

CONTENTS

ACKNOWLEDGMENTS

The solitary nature of typing at a computer screen, sifting through archives, sched-uling interviews, or scribbling down fieldnotes masks the innumerable encounters and underlying structures that make research possible. I like to imagine excavat-ing them all and holding them up to the light, like some kind of delicate botani-cal object, to appreciate the mundane beauty of their intricacies. Doing so would inevitably reveal the incompleteness of the exercise. The closer you look, the more infinite our interdependence becomes.

I suppose as good a starting point as any would be the almond tree. It really took me for a ride. Who knows what I might have written a book about, if at all, had this arboreal figure not rooted itself deep into my consciousness? What people, places, and ideas might I never have come across? I am grateful for the serendipitous combination of time, place, and initial dead ends that eventually drew me to this project. I don't think happenstance and material charisma get enough credit.

I'd also like to appreciate upfront the many types of labor that made my labor possible: the academic staff who supported everyday operations, custodians who kept spaces clean, medical practitioners and food producers who kept me healthy, and childcare providers who made my writing practicable.

I recognize that the land where much of this work took place was located on the unceded territory of Indigenous peoples, including the Awaswas-speaking Uypi Tribe, now represented by the Amah Mutsun Tribal Band, and the Seneca Nation, a member of the Haudenosaunee Six Nations Confederacy. I take that fact as a call to ongoing responsibility.

When it comes to the specific content of this book, I have immense gratitude for the 187 people who sat down for interviews. Sometimes for twenty minutes, but often for hours, they shared their time with no expectation of a direct benefit beyond the satisfaction of telling their story. I want to extend special thanks to those who went out of their comfort zone to talk with me. I hope that everyone who participated can see how they've contributed to an ongoing conversation about how agricultural knowledge operates in our world.

Archivists and librarians are the unsung heroes of so much scholarship. I'm particularly appreciative of the commitment to preservation and access demonstrated by those at the University of California Davis Library Archives and Special Collections; Vacaville Museum; Paso Robles Historical Society; and Madrid's Biblioteca de Agricultura, Pesca y Alimentación.

Funding and material support are essential to this type of work. Thank you to the committees that entrusted me with resources to bring this project to fruition, most notably the Wenner-Gren Foundation and Annie's Homegrown Sustainable Agriculture Scholarship Program, as well as the David Gaines Award and Heller Agroecology Grant based at the University of California Santa Cruz. I could never have buzzed around to so many farms and offices on a shoestring budget without the generous donation of a 2003 Toyota Corolla by my magnificent mechanic brother (it's still running!) or the patience of my partner in teaching me to drive a manual transmission when we got to Spain.

I've been quite lucky on many fronts, not the least of which was to spend my doctoral training in an intellectual culture that fostered lively, creative, rigorous, radical, and humble ways of approaching academia. The University of California Santa Cruz, really has something special. The Agri-Food Working Group, the Science & Justice Research Center, and the short-lived but vibrant Problem of California group were key spaces for developing my ideas. I realize more and more what an incredible gift it was to be surrounded by careful thinkers who were also models of scholarly generosity. Madeleine Fairbairn, Julie Guthman, Carol Shennan, and Andrew Matthews, among many others, set a high bar. There was a time when I was on the fence about writing this as a book at all, and Madeleine's relentless confidence in me pushed me over the edge. As the editor for the Critical Environments series, as well as a mentor and collaborator, Julie's feedback on this manuscript was characteristically thorough, and the book has come out much stronger for it. I'd also like to appreciate the indirect support that came from the academic coven that formed around the UC AFTeR Project, a California Agri-Food Technology Research Collaboration, which included Julie and Madeleine along with Charlotte Biltekoff, Kathryn De Master, Zenia Kish, Michaelanne Butler, and Summer Sullivan.

A special shout-out is due to Emma McDonell and her co-organizers Sarah Osterhoudt and Richard Wilk, whose workshop Critical Approaches to Superfoods at Indiana University in 2019 inspired me to dig deeper into the historical trajectory of almond advertising than I might have otherwise. The insights from that research fundamentally changed the way I understood the industry.

At the University at Buffalo, I've been fortunate to find myself among enormously kind and supportive colleagues. I joined as one of the founding members of the Department of Environment and Sustainability, and it has been wonderful to build something together. Thank you to the students for tolerating all the nut puns and serving as a testing ground for some of the arguments made here. I'm especially grateful for the intellectual community provided by the Critical Ecologies Research Collaborative, including Aniket Aga, Holly Buck, Jordan Fox, Jaume Franquesa, Trina Hamilton, Kirk Jalbert, Marisa Manheim, Lourdes Vera, and Marion Werner. The group workshopped the plant-breeding chapter of this book. The irrigation chapter was workshopped by the Science and Technology Studies Food and Agriculture Network (STSFAN), a fabulously curious and constructive collection of minds that has been meeting monthly via Zoom for nearly four years.

My editor Naja Pulliam Collins and the team at UC Press have been a pleasure to work with. I'd also like to extend my appreciation to the anonymous reviewers for their time and care. Massive thanks to the press for providing the funds to make this book fully open access.

Thank you to the countless friends who showed up for me along this journey, from veggie dinners to Casa Dufour to the YAMs and beyond; you know who you are. Thank you to my parents and siblings for being the kind of family that knows how to dream but doesn't take themselves too seriously. And to Adam and Maya, you enrich my day-to-day existence immeasurably. Thank you for giving me so many superb reasons not to work too much.

Introduction

Naturalized Extraction and Knowing Otherwise

When it comes to the environmental impacts of agriculture, almonds have touched a nerve. As California entered one of its increasingly frequent and severe droughts, back in 2014, almond orchards were not only slurping down roughly 12 percent of the state's managed water supply, they were also expanding across the landscape like never before. The notion that such a "thirsty"[1] crop would be increasingly grown in a "desert"[2] amid a drought baffled onlookers. And the problem wasn't only water. There was also the matter of bees. Depending on who you ask, between 70 and 90 percent of commercially managed honeybee hives across the entire continental United States are being trucked long distances and exposed to all manner of agrichemicals and contagious bee ailments in order to pollinate almond orchards each year despite a spiraling pollinator health crisis. Given the industry standard of two hives per acre, almond orchards "require" more beehives than are available in the entire continental United States. Deeply dependent on insect pollination, almond growers have been willing to pay unprecedented sums to lure beekeepers out west knowing the hazards it might entail. During the drought, almond orchards were more profitable than ever, the state's most valuable farm export by far, and new plantings showed no signs of slowing down. For those concerned with the social and ecological ramifications of industrial farming, the logic followed that almonds were simply too needy, too thirsty, too bee-dependent, and too out of place in California's semiarid climate to be sustainable.

Imagine my surprise when I found that the polar opposite could also be true. In Spain, the world's largest producer of almonds until the 1970s, almonds have long been known as the most rugged, resilient, undemanding of crops, uniquely capable of producing in harsh environments. The vast majority of the country's almond growing landscapes use no irrigation. In these rainfed orchards pesticide use is considered largely unnecessary. Hauling honeybee hives to almonds for pollination is almost unheard of. Often called a crop of convenience, almonds epitomize a low-maintenance land use for those with a rough patch of earth and little time and money to spend on it. Farmers celebrate almonds for their ability to produce despite intense heat, drought, steep slopes, and low-nutrient soils. Understandably, yields are much lower under rainfed and low-input conditions, often about a third of that of their irrigated counterparts. Spanish farmers are not performing some kind of miracle, and eventually the record prices of 2014 provoked a surge in California-style plantings. But what I struggled to understand was how such opposing characterizations of a single crop could both be true. How could almonds embody the epitome of thirst and of thrift? Of neediness and self-sufficiency? To take the question a step further, how does knowledge about a given crop's "needs," and thus its relationship to the surrounding environment, come into being?

This is the crux of the almond paradox. My concern is not with almonds specifically but with what this contrast can reveal about the politics of agricultural knowledge. This book is an effort to challenge the taken-for-granted claims about a given crop's requirements—the quantity of water or pollinators it needs, the characteristics it needs to exhibit, and the places it needs to be grown—and denaturalize them. These statements are not inherent characteristics of the plant's physiology. As I will detail, they reveal quite a lot more about the entanglements of capitalist enterprise, the state, social hierarchy, and scientific practice than they do about plants.

Take the almond's "crop water requirement" for example. The equation for calculating this amount used by agronomists around the world currently spits out a number that surpasses the amount of rain falling anywhere on Earth where the climate permits almonds to grow. By this logic, an almond tree "needs" more water than its environment could ever provide. A seemingly neutral description of plant-water relations naturalizes extraction, from elsewhere or underground, generating profits at the expense of ecological integrity. My point here is not to chuck out agronomic knowledge processes as meaningless but to insist that they are always political, no matter how banal an equation for calculating irrigation applications may seem. Forging an agricultural future that invests in social and ecological thriving requires reckoning with the knowledge systems we have inherited and reimagining them.

Unpacking the political underpinnings of what plants need is not merely an academic exercise. There are several key implications that follow from denaturalizing agronomic claims. One is to recognize that knowledge about farming, even the most mundane, is inextricably intertwined with questions of land, labor, power, and profit. Such a claim is familiar within the field of science and technology studies, but it rarely makes its way into conversations about agriculture. Synthetic pesticide and fertilizer dependence has received a good deal of attention, given its clear departure from a "natural" state of affairs,[3] but agriculture's relationship to water, pollinators, or particular geographic conditions risks being taken at face value. Claims about these relations are much less static or inert than they seem. Popular claims, like the crop water requirement, are more than naive and ecologically negligent, a relic of a time before scientists understood the true scale and scope of our predicament. It is far more illuminating to recognize these knowledge practices as part and parcel of a system predicated on extraction and thus gain clarity as to what it might take to operate otherwise.

When I use the term *extraction*, I refer to a logic in which matter (both living and nonliving) is manipulated in ways that generate short-term profit while undermining longevity or renewal. Extraction here does not refer to a particular set of material processes, such as the removal of groundwater or minerals, but rather to a type of parasitic relation. This lopsided dynamic is inherently unstable, as capitalist enterprises rely on the very material processes they degrade.[4] Extraction is also a matter of justice, as it is fundamentally predicated on the notion that certain landscapes or bodies are more disposable than others.[5] Extraction has been a core feature of settler colonialism and its severing of eco-social relations.[6] What unsettled me about the almond case was the way in which claims about a plant's needs allowed extractive relations to be described in botanical terms. It gave the impression that extraction was simply a fact of nature.

One key mechanism for allowing extractive relations to become naturalized is the presumed placelessness of agronomic information, especially where efficiency is concerned. When it comes to environmental impacts, efficiency has become the guiding star for agricultural industries. A prominent 2018 report by the World Economic Forum and McKinsey & Company positioned efficiency as the key to agriculture's sustainability improvements in the decade to come.[7] The United Nations has made "increasing resource efficiency" a centerpiece of its Environment Program.[8] CropLife International, the most powerful global agribusiness industry group, claims that achieving food security while preserving biodiversity requires "increasing productivity and efficiency" and lobbies policymakers to adopt this vision of sustainability.[9] During the 2014 drought, the California Almond Board proudly touted that their orchards were producing each pound of almonds with a third less water than they had twenty years prior.[10] That sounds like good news. But when I dug a little deeper, I found that for every acre of almonds, water use on that patch of land had actually increased by 37.4 percent. Yield boosts due to density and agrichemical use masked the fact that irrigation had increased dramatically. In a place with falling and increasingly contaminated water tables, with land subsiding in places as much as a foot per year,[11] calling this a victory for water conservation works only if you have very powerful blinders on. Making the politics of agricultural knowledge explicit means not only calling out the absurdity of such blinders but also asking how the blinders were put on in the first place and why it's so hard to take them off. In this book, I aim to show how agronomic knowledge is inextricable from the configuration of capital, scientific institutions, and the state and from the water, air, land, and organisms through which they operate. Imagining agronomic knowledge to be untethered from the specificities of its socioecological context is not only inaccurate, it is dangerous.

Efficiency acts as a kind of antiplace, a way of sidestepping recognition of the immediate ecological context of the farm. It also plays right into the hands of one of the biggest culprits of industrial agriculture's environmental hazards: over-production. As agronomists strive to make orchards more efficient, they strain the market's ability to absorb enormous volumes of what used to be quite a niche product. This book charts how the almond industry's boom, and perpetual fear of bust, serves to shape knowledge practices. The marketing-driven profitability the industry saw in the 2010s not only sparked a global planting boom but also fundamentally challenged the way almonds were understood in places like Spain with long-standing rainfed production systems. This has risks. On the one hand, high prices have given regions with effectively organic low-input production systems hopes for rural revival. On the other hand, these hopes may be short-lived as production swells and prices predictably fall. The export of California-style agronomic recipes for intensive almond production have made the market far less hospitable to rainfed growers. Moreover, the boom has challenged the persistence of the almond tree's historically complimentary role in a diversified agricultural system. When marginality becomes seen as a defect, rather than an asset, eco-logically and socially desirable landscapes can fall into abandonment while sites of acute socioecological precarity continue an extractive, downward spiral. In other words, placeless agronomic knowledge accelerates the almond boom's social and environmental harms while risking the displacement of the very kinds of holistic and place-based practices that offer long-term viability.

CALIFORNIA'S ANTICIPATORY OVERPRODUCTION

Perhaps the most striking thing about the environmental toll of almond intensifica-tion is the fact that it seems so wholly unnecessary from the perspective of meeting consumer demand. By and large, people have not been begging for more almonds. Quite the opposite. Until the early twentieth century, almonds were considered a seasonal food, typically eaten in association with winter holidays.[12] In the United States, grocers routinely refused to stock almonds for the first eight months of the year.[13] It took nearly four decades of industry-funded publicity to overcome this cultural association and convince people that almonds could be eaten casually in meals or snacks regardless of the season.[14] It would be as if chestnut growers began an epic campaign to promote the consumption of roasted chestnuts (typically a winter novelty) not only year-round but also with breakfast, lunch, and dinner. One could make the argument that such a chestnut-enriched diet might be nutri-tionally or even ethically desirable, but that's beside the point. The point is that the almond industry's growth has been very clearly driven by growers' anxieties over farm profitability, and nearly all signals of consumer demand can be attributed to the industry's promotional efforts. So much of the pesticides, the water, the soil contamination, the pollinator entanglements, the risk of losing ecologically sound production systems are an unfortunate byproduct of California growers' perpetual race against time to avoid a price crash. The industry has essentially been a victim of its own success. The more adept almonds growers were in maintaining profits despite rising production, the more farms converted to almonds and spurred a fresh round of overproduction panic.

Like other capitalist fruit growers, California's twentieth-century almond growers exemplified the motto "organize and advertise."[15] In 1910, after a long series of negotiations, more than 80 percent of all almond production in California (and thus the United States) came under the marketing umbrella of a single entity, the California Almond Growers Exchange (CAGE). This collective bargaining power drove prices up by 50 percent in the decade to follow, and many more growers decided to get in on the almond game. California's agricultural history distinguishes itself from most other settler-colonial contexts in being thoroughly capitalist from its inception, rather than populated by self-sufficient homesteaders.[16] Agriculture served as a place to sink rapidly accumulating capital from mining and related industries. The 1849 Gold Rush and ensuing claim to US statehood had triggered the violent theft of land from Indigenous communities and legal trickery to oust Mexican settler ranchers, expropriating vast swaths of land for investment.[17] The Swamp Lands Acts passed soon after statehood awarded parcels to those who would drain and "civilize" wetlands, an economic and racial project to attract white settlement that radically transformed the ecology of the Central Valley.[18] Almonds initially became a popular investment crop in the Sacramento delta and river valley, as well as the central coast near Paso Robles.[19]

The exuberant expansion of almond orchards by gentleman farmers, and sometimes far-off investors, at the turn of the century quickly saturated regional markets. European almonds, primarily from Spain, held distinct advantages for buyers in population centers of the eastern United States: they came preshelled (a labor-intensive process that California growers could not afford) and shipped more cheaply by sea. By 1919, the fledgling California almond industry faced a crop twice the size of the year prior. Keep in mind, at this time almonds were rarely if ever irrigated, received minimal organic fertilizer and pesticides, and did not rely on any imported honeybees, but overproduction was already a serious menace. The Growers' Exchange president warned farmers, "You will have much to worry about . . . if you fail to supply the necessary funds for advertising and development."[20] He explained, "The consumer will consume only to the extent that you create a demand by educating him in the value and attractiveness of your product."[21] And boy did they ever.

In the Roaring Twenties, almond marketers distributed elaborate promotional displays for grocery storefronts. When the Great Depression hit, the industry hired nutrition science consultants to validate health claims that were disseminated along with recipes through print media, radio, cooking schools, nutrition educators, hospital dietitians, and military quartermasters.[22] When the United States entered the Second World War, CAGE successfully lobbied for almonds to be designated an "essential food" by the War Manpower Commission. When the war ended, they landed an agreement with the USDA school lunch program to buy five million pounds per year of almonds. They also began offloading almond surpluses abroad at half price to drum up new markets.[23] But postwar technologies made matters worse. Synthetic fertilizer, pesticides, mechanization, and more intensive irrigation sent yields soaring 64 percent between 1949 and 1961, prompting the launch of the Colossal Almond Crop promotional campaign. The governor of California, in consultation with the industry, even considered resorting to legal restrictions on almond plantings in 1966 but determined market creation was more politically practicable. Then California almond growers, dreading oversupply, did something truly unprecedented.

In 1972 California growers joined other commodity groups in convincing Congress to unlock a massive flow of cash for food advertising. Federal marketing orders, a legal mechanism to steady farm incomes created in the wake of the Great Depression, allows farmers to form commodity groups (in this case the Almond Control Board) with the power to collect a fee per pound from all processors. This "assessment," essentially a tax passed on to farmers, funded price-stabilization tactics such as stockpiling surpluses and setting grade standards, but advertising was always strictly off limits. Then suddenly, like the flick of a switch, this prohibition was removed. Just a few years later the Almond Control Board spent 97 percent of its budget on advertising, promotion, and research and development.[24] The promotional challenge they faced was formidable. Already struggling to sell mounting stockpiles, but managing to keep prices reliable, almond acreage during the 1970s doubled. Backed by the marketing order's new funding capabilities, CAGE (rebranded as Blue Diamond in 1980) literally pleaded with the American public. A television ad featuring farmers up to their elbows in almonds begged viewers, "A can a week is all we ask."

The ability to charge producers a fee that is then deployed for advertising has been wildly successful.[25] Perhaps too successful for the industry's own good. Health-oriented marketing, bolstered by industry-funded nutrition research and a "qualified" endorsement by the American Heart Association after heavy lobbying, yielded strong sales in the 1990s.[26] New products like almond milk, crackers, and cosmetics proliferated to absorb mounting production. The Almond Board's ability to consistently evade a price crunch with its stunning marketing savvy attracted more and more growers struggling with surpluses in other commodities, like cotton and tomatoes, to plant almonds. Meanwhile university researchers developed methods for greater production efficiencies pumping out more almonds per acre. The boom hit its zenith around 2014 and proceeded through the state's historic drought, shockingly unconstrained because it was still the best bet for farmers and a growing cohort of farmland investors,[27] generating nearly $8,000 in profits per acre on average.[28] Volumes ballooned as new acreage was compounded by intensified practices that had doubled yields per acre over a twenty-year period. By 2016 the Almond Board was so panicked about the increase in plantings that it sought approval from the USDA to increase the assessment by 33 percent (from three to four cents per pound) for three years in order to swell its advertising budget.[29] Even with this enormous spending capacity, prices fell by more than half, back to the level of the early 2000s. In 2023, new plantings slowed considerably, but they have yet to stop.

Overproduction is typically described as reactionary; a glut of corn or wheat pushes prices so low that farmers have no choice but to increase their production efficiency or get out of the game.[30] Production efficiency can come from lowering input costs, but more often it comes from capital-intensive technologies to boost yields, like irrigation, agrichemicals, machinery and even imported pollination, as discussed in the following chapters. The technologies deployed to save the farm from bankruptcy, or even make a pretty penny now and again, incur steep environmental costs. Nearly all industrial commodities follow this pattern. Corn is perhaps the most well-studied case of a commodity surplus desperately searching for a market as the technological treadmill pushes yields higher and profits lower, forcing many smaller farms into bankruptcy. What is interesting about the almond case is that the same kind of treadmill has been forcefully underway *in anticipation* of a price crash, which appears always on the horizon. In addition to the typical suite of agrichemical inputs, marketing serves as yet another "input" of sorts, temporarily alleviating but ultimately deepening the looming overproduction crisis.

Given this context, it is quite odd that the popular press placed much of the blame for the intensity and expansion of almond orchards on consumers. "Lay off the almond milk, you ignorant hipsters," a provocative 2014 *Mother Jones* article griped.[31] "People are eating almonds in unprecedented amounts. Is that okay?" *The Atlantic* queried readers.[32] "China's taste for almonds is sucking drought-stricken California dry," the Associated Press melodramatically proclaimed.[33] It's true that by 2016 Americans were eating, on average, four times as many almonds per capita as they had twenty years prior.[34] The almond industry had worked hard to drive demand while receiving a serendipitous boost from food trends toward high-protein, plant-based, and gluten-free diets. And it's true that the growing middle classes in India and China, where the Almond Board has invested enormous sums in advertising, have become vital almond export markets. But blaming consumers in the United States or abroad for demanding so many almonds is almost like blaming consumers for demanding so much high-fructose corn syrup. Nobody asked for this.

SPANISH ALMONDS AT THE MARGINS

Until the mid-1970s, right when the California almond industry unleashed its publicity prowess on the world using producer fees, Spain was the world's number one almond producer. Orchards were almost universally rainfed, using no pumped or diverted water whatsoever. Fertilizer, if used at all, typically came from sheep manure in adjacent pastures. Pesticide use was negligible. Fossil fuel use was limited, as a tractor might be used for shallow plowing but the harvest was done by hand. It was an extensive low-input, low-output system. One that had extended so widely across Spain's rural landscapes as to achieve a place of global dominance almost effortlessly.

The explanation for this largely comes down to marginality. While agrarian capitalism could be as fierce and extractive in Spain as anywhere, almonds remained persistently at the edges both literally and figuratively. Known for being rugged, resilient, low-maintenance trees, almonds have historically been planted at the fringes of agricultural viability. They serve as borders between fields, shade along roadsides or amid pastures, and erosion control along steep slopes or acequias (gravity-fed water canals). They work well as a complement to staple commodities like wheat, wine grapes, and olives. Often their maintenance is considered secondary, an extra opportunity once the real work is done. The yields from almond trees swing wildly from year to year, given sensitivity to rain and frosts, making them quite unreliable for base income but ideal as companion plantings. Where almond trees do appear as a primary crop, landowners typically own a small parcel of low-quality land and earn their living with wage work elsewhere. Throughout Spain I heard the same refrain from farmers, word for word, as if it had been collectively rehearsed: Almonds are what you grow where you can't grow anything else.

A question worth asking then is, How is it that a crop of last resort became so ubiquitous? It is true that Spain has quite a bit of land considered marginal for agricultural purposes: steep slopes, low-nutrient soils, and low rainfall. But vineyards and olive orchards can often perform well under these conditions. What is perhaps more significant in explaining the almond phenomenon is the way these trees have become intertwined with social marginality. In nearly every form rainfed almonds take in Spain, their arrangements reflect stark rural inequities, the instabilities of agrarian capitalism, and the resourcefulness of those struggling for a dignified livelihood. If rainfed almonds are known among Spanish farmers as a "crop of the poor," then their abundance does not reflect the workings of a romanticized, ecologically attuned peasantry so much as the pervasiveness of rural hardship. The vast and diverse landscapes on which almonds grow are also a result of overproduction in other commodities, like wine grapes and olives, and the perpetual pursuit of fixes to capitalist crises. Much like California, agricultural intensification has wreaked havoc on Spain's ecosystems, with reengineered rivers, depleted aquifers, toxic agrichemicals, and chronic nitrogen contamination. Yet somehow, until the boom of the 2010s, Spain's almond ecologies managed to escape the technological treadmill and avoid the extractivism fanning out all around them. By some miracle, almonds appeared to be immune.

Almond trees were likely introduced to the Iberian Peninsula during the Greek or Roman Empires, with origins in the dry mountainous regions of Central Asia.[35] In regions along the Mediterranean coast, they served as a compliment to the trio of wheat, olives, and vineyards. Certain areas with particularly challenging agricultural conditions specialized in almonds as early as the sixteenth century, such as Jijona, now famed for the almond-based dessert turrón.[36] It's difficult to know exactly how significant almonds have been historically for farming households, in most cases dedicated for local consumption, but their commercial importance seems to have remained minor and isolated to niches within coastal regions until the late nineteenth century.[37] When the phylloxera wine blight began tearing across France in the 1860s, grape prices spiked and a planting frenzy ensued in Spain, which remained protected for several decades by the Pyrenees mountain range. Older mixed orchards including almonds were torn out to make way for lucrative vineyards.[38] But the good luck didn't last long, and when phylloxera finally did make its way to Spain, almond trees were one of the popular replacements along with olive and carob.

Almonds steadily gained ground in Spain around the turn of the twentieth century amid a period of widespread agrarian crisis in Europe. Despite protection measures, imported wheat from around the world drove down prices, wine production surged in French-controlled Algeria, olive oil faced competing products in global markets, and shipping was cheaper than ever as even Spanish livestock lost its edge to Argentina and New Zealand.[39] Many large landowners saw that agriculture was losing its prominence as a site of wealth and sold off lower-quality parcels in order to shift their investments into industry.[40] These small plots at the fringes of agricultural viability were often incapable of sustaining a household, and almonds became a crop of choice for those needing a low-maintenance land use while they worked for a wage.[41] As an agronomist in 1912 admired, "the advantages of this precious tree are incalculable, which grows in the poorest soils, almost unsuitable for any other crop, and withstands the frequent droughts of our peninsula" without complaint.[42] Spanish farmers I interviewed, most of whom were born in the 1940s and 1950s, often explained that their grandparents' proudest achievement had been managing to save enough as day laborers to finally buy a patch of land. Almonds had not been anyone's full-time job. They were one piece of a complex livelihood strategy that provided not only income in good years but also a sense of stability, pride, and hope for the next generation.

Spain's agricultural economy varies widely by region, but it has generally been characterized by the pervasiveness of two land-use archetypes: the large estates worked by landless laborers, known as *latifundia*, in the south and the densely fragmented parcels barely big enough for subsistence, known as *minifundia*, in the north. Under both systems, almond trees served a kind of ecosocial niche, finding their way into the nooks and crannies of agricultural landscapes largely as a coping strategy for those enduring substantial precarity. Rural class conflict was an underlying driver of the Spanish Civil War of 1936–39, which came on the heels of significant land redistribution efforts by the Republican government.[43] When the fascist dictator Francisco Franco took power, his agricultural policies emphasized autarky, incentivizing domestic production of staple foods, especially grains, and limiting participation in international markets (which were already shunning Spain due to its alignment with the losing parties of the Second World War). This approach led to hunger, a booming black market, and desperation-driven migration to cities. Franco thus began softening his hard-line economic tactics and opening Spanish agriculture to international trade. As part of this incremental opening, and under pressure to increase productivity, his government set out in 1951 to increase the "reforestation" of the countryside, by incorporating almonds (along with carob, figs, olives, and vines) on land that was "inappropriate for crops of other classes due to its quality, topography, or clear risk of erosion."[44] The policy reflects a growing market for almonds, due in no small part to California growers' global marketing efforts, while at the same time formally asserting the almond tree's place at the margins of agricultural viability.

Almond prices rose throughout the 1960s as overproduction pushed down the prices of wine and olive oil, and Spanish growers took note. In a single year (between 1970 and 1971), land dedicated to almonds jumped by 24 percent, and by the end of the decade it had practically doubled. Production once confined to small plots and hillsides began to stretch over larger areas where grapes and olives had once been, allowing a scale and slope suitable for mechanized tillage. At this point, Spanish agronomists started to take a more sustained interest in "modernizing" almond production. Whereas a 1963 agronomic guide describes common ways almonds can be interplanted with grain rotations,[45] its 1970 counterpart describes this diversified system as prevalent but "catastrophic" and insists on monocultural specialization.[46] The planting spree tapered off by the 1980s. Agronomists griped about the persistence of "irregular" plantings, sheep grazing in the understory, which limited plowing, and the resistance of farmers to using yield-boosting agrichemicals. Yet they simultaneously marveled at the fact that in certain regions, "despite its low productivity, it's still one of the most important contributors to the island [of Mallorca]'s GDP."[47] This combination of ubiquity and low productivity was largely attributed to the fact that most almond-growing households did not depend solely or primarily on almonds and had limited purchasing power. Spain's rapid urbanization during the "economic miracle" of 1959–73 also meant that many rural properties no longer had full-time residents and incomes were much more strongly tied to the construction and tourism industries.[48] Almond trees were, in today's lexicon, a side hustle: a secondary income strategy used to compensate for low or unstable pay. They were also, for many urbanizing families, a low-effort means of maintaining connection to their rural roots.

California's treadmill of overproduction, when farmers up to their elbows in almonds were pleading with consumers to buy a can a week, was certainly a source of worry for Spanish growers. But never quite enough of a worry to prompt small-scale operations to make hefty investments in industrializing their orchards. In 1986, a democratic post-Franco Spain joined the European Union, and the nut industry was targeted by one of a raft of new export-oriented subsidy schemes to correct this perceived problem. As part of a 1989 law establishing funds to support "improvement plans," almond growers could receive payments for tearing out old trees and planting new varietals, applying fertilizer and pesticide treatments, enhancing pollination, and acquiring technical assistance. Beneficiaries were required to become members of a local producer organization and to make their fields "homogeneous," without scattered trees.[49] While many farmers took advantage of the program, especially to fund new plantings of late-blooming varieties (see chapter 3), a decade later it had made no appreciable impact on average yields and prices were tumbling to new lows. In the early 2000s, Spain's expansive rainfed almond landscapes began to shrink for the first time in at least a century.

Unbeknownst to most Spanish farmers, the Almond Board of California was deploying its ballooning marketing budget to muscle its way into new markets, new product categories, and new culinary trends. And it was working. Demand grew even faster than expected. When drought hit California in 2014–16, prices hit unthinkable highs, despite the fact that overall almond production remained stable, with overall yields well above the five-year average predrought.[50] Spain's rainfed regions saw an uptick in almond plantings, but a profound paradigm shift was underway. Suddenly almonds were no longer a crop of convenience for rural households squeaking by but instead the darling of investment groups, irrigated districts plagued by sinking commodity prices, and industrial-scale agribusinesses. Many agronomists I met described how almonds had been transformed into a fundamentally new kind of crop. No longer "marginal," they were now a "frutal" (a fruit tree). High capital, intensively irrigated, chemically dependent, California-style orchards were sprouting up in prime agricultural land, radically shifting not only how but also where almonds were grown. While still a small minority by land area, these intensive orchards would become the majority of Spain's production in 2022.[51] Even for those households maintaining rainfed almonds, the industry will never be the same. The boom marked not only a tipping point in how Spain's almonds are grown but a revolution in how they are known.

NATURALIZING EXTRACTION

This book compares almond ecologies in California and Spain to ask a simple question: How can a plant, and its relationship to its surroundings, come to be known in radically different ways? To answer this question, I focus on the notion of needs. How much water or pollination does an almond tree need? What characteristics does it need to exhibit, and where does it need to be? Tracing the dual histories of breeding, irrigation, pollination, and place across these geographies reveals how political economic dynamics have been fundamental to knowing what plants need. I'm interested in the ways that this form of agronomic knowledge generates a naturalizing effect, masking and perpetuating the extractive relations it describes.

The agronomic claims surrounding almonds are emblematic of the blurry boundaries between the biophysical sciences and political economic knowledge practices. This conceptual slippage can come in many forms, as often highlighted by scholars in the interdisciplinary field of political ecology.[52] The concept of carrying capacity, for example, refers to the number of individuals of a certain species that a given environment can sustain. The idea originated as a method for taxing cargo ships and was later applied widely to managing grazing animals on rangelands. Despite serious shortcomings as an ecological model, carrying capacity was highly effective in facilitating banking and government land sales.[53] Its persistence is reflective of its social context, yet these relations are masked by notions of scientific objectivity. Similar dynamics can be seen in the way neoliberal economic policy has come to pervade environmental sciences, with countless studies reframing biophysical processes as ecosystem services.[54] The economy itself is even represented in ways that suggest it is an organism following an evolutionary trajectory.[55] Naturalness is a powerful rhetorical tool. And even though scientists may be aware of the socially situated nature of their work, at some point it no longer seems worth mentioning. Political economic pressures on farming and ecosystems become so hegemonic as to be unremarkable or even presumed inevitable.[56]

In the case of knowing what plants need, complex political economic relations are woven into scientific practices and then quickly invisibilized through the language of ostensibly inherent physiological qualities. It's like a disappearing act. What I want to explore here is the process by which economic conditions become rendered as ecological, and the work that this naturalizing effect does. Placelessness, I argue, is a central mechanism for invizibilizing politics. The consequence is substantial precarity, both for the people and places intertwined with extractive production as well as those holding on to more sustainable alternatives. I see these insights as a contribution to conversations at the intersection of feminist science studies and agrarian political economy.

Feminist science studies is a body of scholarship that broadly examines knowledge processes as sites of power relations.[57] This type of analysis makes the social fabric of scientific practice explicit. It challenges objectivity and universalism, insisting that knowledge is partial, situated, and inescapably political.[58] A central strand of this work investigates contestations around how scientific claims are developed, defended, and contested, such as germ theory, radioactivity, chemical toxicities, and colony collapse disorder.[59] Researchers most often ask how a concept or claim becomes ascendent and maintains durability or how groups navigate scientific uncertainty.[60] In this book, rather than controversy or contestation, I'm interested in unquestioned and seemingly inconsequential forms of basic agricultural information. My argument here is not that claims about what plants need are inaccurate but rather that they are doing active political work.

This project builds on the momentum of scholarship across geography, sociology, and history of science, progressively unpacking the politics of agricultural knowledge production. Jack Kloppenburg's *First the Seed* traces how genetic engineering was advanced largely to secure the dominance of corporate agribusiness interests.[61] Deborah Fitzgerald's *Every Farm a Factory* details how industrial manufacturing became a template for modern agronomy, converting farmers into business managers.[62] Susanne Freidberg's two monographs explain how colonial relations continue to shape food quality perceptions,[63] and how the modern preoccupation with freshness was produced through globalizing markets.[64] Christopher Henke's *Cultivating Science, Harvesting Power* conceptualizes agricultural extension workers as serving to temporarily "repair" industrializations' many vulnerabilities.[65] Julie Guthman's *Wilted* emphasizes the structural pressures on scientific practice that leave many promising paths toward resilience unstudied.[66] Most recently, Kelley Bronson's *The Immaculate Conception of Data* exposes the ways in which digital agricultural information can reproduce long-standing inequities.[67]

In *The Almond Paradox*, I scrutinize and denaturalize one of the most fundamental and mundane elements of modern agriculture: understanding what plants need. I aim to unsettle the way in which these claims are often treated as stable truths rather than evidence of escalating addictions. Perhaps more than most authors listed here, my approach is also rooted in material semiotics. By this I mean that I share the concern that social scientists have for too long set aside the material aspects of plants, water, insects, air, soil, animals, and human bodies or any other beings in our work, presuming such matters are beyond our intellectual jurisdiction or underestimating their significance. A desire to extend the "social" to encompass relations among all manner of beings, and to dethrone humanity's exclusive claim to agency, is at the heart of scholarship characterized as new materialist, posthumanist, multispecies, and more-than-human.[68] Such approaches, however, have often spent so much effort animating a plurality of actors that they inadequately address questions of power, intersectional inequality, settler colonialism, and critical political economy.[69] I found material semiotics, which insists on the inseparability of matter and meaning, to be a useful tool for sustaining the fusion of these theoretical, political, and ethical commitments.[70] I also came to understand the propensity toward depoliticization within some strands of the more-than-human turn as a symptom of its lack of engagement with Indigenous thinkers who have long centered kinship and reciprocity among all beings.[71] As postcolonial scholar Syliva Wynter points out, the very definition of what we call human has long upheld the interests of a small portion of the world's population.[72]

TREES ON A TREADMILL

I began this project eager to understand how power relations become baked into plant knowledge, but that's not all that I came away with. The almond industry's particularities also led me to rethink some core assumptions about the political economy of agriculture and its relation to technological change. The technological treadmill, as theorized by Willard Cochrane, explains that farmers are effectively compelled to adopt new practices by falling prices.[73] A new seed variety or irrigation strategy or chemical input increases profits for early adopters, but once it becomes commonplace, overall production of the commodity rises and prices fall. A "laggard" farmer facing low prices can either increase their yields with new technology, scale up their operation, or risk going out of business. The almond case led me to rethink the temporality of this process and to expand what typically counts as technology.

Almonds, and likely other permanent crops like vines and other trees, exhibit hypersensitivity to the *risk* of a price fall, even if it never occurs. This means technological change is incentivized without the actual experience of a price drop and is rooted more in market anxiety than reality. The upfront investment for an orchard is quite high, as growers purchase costly saplings rather than seeds and supply trees with inputs for several years before their first harvest. They also expect the trees to live twenty years or more. Thus the risk of a price fall is not just of losing profits from one year but from several years in the past and potentially many more in the future. The longer time horizon of orchard investments also means that a boom in production can be calculated many years in advance, heightening fears and motivating preemptive measures. The Almond Board of California keeps careful measurements of new plantings and nursery sales in order to estimate production three, five, and ten years down the road. Thus the suite of technologies for increasing yields (intensive irrigation, pest control, and pollination), cutting costs (mechanization and reduced pruning), and supercharging demand (advertising and export) are attractive long before farmers feel the sting of a price fall. Permanent crops get on the treadmill even when their present circumstances seem rosy because the threat of overproduction is both scarier and easier to see swelling on the horizon.

The concept of the technological treadmill centers decision-making by individual growers and the tools they deploy on the farm. What I see in the almond case is the immense influence of, and addiction to, a collective technology: marketing. Steven Stoll's landmark study of fruit growers' "organize and advertise" strategy focuses on marketing as a solution to the spatial challenge for California farmers in reaching the majority of the US population living thousands of miles away.[74] The almond case demonstrates that not only are these marketing strategies crucial to overcoming distribution challenges, they are also an ongoing requirement to stave off an anticipated price fall. I see marketing as a technology, just like agrichemicals or machinery, upon which almond growers became dependent. As with pesticide applications, which often increase due to diminishing effectiveness, these investments in boosting demand tend to ratchet up over time as existing markets become saturated and attracting new customers becomes more costly.

State power has fueled marketing's role as a technology for tempering the threat of overproduction. The Almond Board of California is highly effective because, like other federal marketing orders, it has the force of law to collect its dues, overcoming a collective action problem among growers. The strategy, however, has a serious flaw. The almond boom of the early 2000s saw the industry become a victim of its own success. Demand was driven so high, by a combination of deft advertising and serendipitous food trends, that the profitability of almonds attracted many more farmers to plant almonds. Land that had been dedicated to other crops or nonagricultural uses transitioned to almond orchards. This only made for a bigger headache for growers as the enormous surge in production threatened their margins. Marketing is thus an inherently imprecise tool in the technological treadmill that can backfire spectacularly, deepening dependence on all manner of farm practices that promise to boost profits.

These insights are the kind of nuggets that often emerge from the commodity studies tradition. Detailed analyses of the lettuce, sugar, strawberry, milk, cotton, banana, and soy industries, among countless others, have used the market mechanisms and narrative continuity of a commodity to map out the social and ecological contours of agrarian capitalism.[75] Each of these works demonstrates how food or fiber production is tied into extractive relationships to land and labor. The methodological approach is itself a product of industrialization, as its viability rests on the existence of monocultural plantations. There is a risk here of losing sight of all the ways in which deviations from this model persist and resist. Part of my commitment to the comparative work in this book was to keep those tensions squarely in view. Feminist political economists Gibson-Graham challenged scholars to reject totalizing narratives of capitalism, reminding us that we may inadvertently be contributing to the very dominance we seek to critique.[76] Anna Tsing's work tracing relationality with matsutake mushrooms exposes how even in a single object, commodification itself remains patchy and incomplete.[77] It is in this spirit that I hope to provide a more-than-capitalist commodity study that reflects the frank realities of agrarian capitalism and its overwhelming pressures while also doing justice to the otherwise.

ALMONDS TO ALMONDS

Comparison is a complex and inherently imperfect endeavor. Many (though not all) agronomists I met along my research journey seemed perplexed at my project, explaining that almond agriculture in California and Spain is so fundamentally different that comparing them would be like comparing apples to oranges. Like many who study the politics of knowledge, I found that resistance fascinating. After all, in the global market, almonds from these two locales are constantly being judged in relation to one another. How are the boundaries of permissible comparison drawn and why?

For those coming from a positivist approach to research premised on controlled experiments and hypothesis testing, it was easy to see how comparing these two cases with wildly different historical conditions and political economic arrangements would create an unmanageable number of variables. There is no means by which distinct explanatory features could be completely isolated or deduced. I had the hunch, however, that this was not the biggest source of unease with my comparative project. The resistance seemed to come, in part, from an attitude among agronomic professionals in both places that rainfed agriculture is not an equally valid form of knowledge at all but rather one that is uninformed, inefficient, archaic, unscientific, or simply destined to disappear. This deficit model has often been projected onto marginalized groups.[78] It has also validated the status quo much in the same way as using the term *conventional* normalizes industrial agricultural practices. For many Spanish agronomists, trained to maximize production efficiencies, the comparison also provoked a twinge of embarrassment that their farmer clients were not as "professional" as their California counterparts. Another undertone in the resistance to comparison, I suspect, came from a protective reflex, a need to deem these worlds irreconcilably different because acknowledging rainfed agriculture as valid, even useful, would be an admission of complicity in the disastrous consequences of intensification on some level. It is quite uncomfortable for someone trained to teach farmers how to boost plant productivity to acknowledge that just a few decades ago most of the global demand for almonds was satisfied by a secondary side hustle of a crop grown with minimal effort and minimal environmental impact. The comparison quickly does away with the pervasive legitimating narrative that extractive methods are necessary to "feed the world."

Like all research endeavors, what we choose to study and how has political implications. My choice to consider the agronomic decisions of nonspecialized, rainfed farmers on the same level as industrial growers explicitly pushes back on this tendency to devalue nonhegemonic knowledge practices. I find comparison to be extraordinarily helpful in denaturalizing dominant claims. There is more than one way to know a tree. Alternatives to an extractive agricultural treadmill of overproduction are not only imaginable, they are here and now if we look in the right places. Moreover, we don't have to go seeking out ideologically driven social movements to find them. For many decades, work in critical agrarian studies has centered the inspiring efforts of food sovereignty groups like La Via Campesina, agroecology networks like Campesino a Campesino, activists fighting genetic modification, or organizers advancing organic methods, localism, or regenerative practices.[79] This work is invaluable. And yet at times it can appear as if non-extractive farming methods are always a matter of staunch ethical conviction, or associated with poverty in places long exploited by wealthier nations. Part of what struck me about the almond paradox was the way in which it failed to fit neatly into these existing narratives. Rainfed almonds in Spain are not a product of ideological zeal nor a symptom of geopolitical inequity so much as evidence of the messy, patchy, lively, and incomplete nature of agro-industrialism.

California and Spain share an enormous amount in common, and their agrarian histories are more intertwined than one might expect. Both are the metaphorical fruit and vegetable baskets of their respective regions, with agricultural economies centered on intensive horticulture. By value, California produced 62 percent of fruits and nuts and 60 percent of vegetables in the United States in 2022.[80] Spain is the top fruit and vegetable producer in Europe, producing 28 percent of fruit and 21 percent of vegetables by volume that same year, and is the top exporter to countries within and beyond the European Union.[81] Similarly, both California and Spain are leaders in commercial apiculture (honeybee husbandry) in their respective regions.[82] This leadership in horticulture is largely attributed to the long, warm growing season along with, somewhat paradoxically, the factory-like intensive production enabled by a dry climate (see chapter 4). With intensive agriculture centered in regions characterized by extended summer droughts with almost no precipitation, surface waters in both countries have been rerouted on an extraordinary scale. Elaborate networks of dams and diversions effectively plumb their entire hydrologic systems. Spain's mammoth public water-moving projects, built under Franco's technocratic regime, actually took California's early era of dam building as their inspiration.[83] Early twentieth-century leadership in both regions viewed the contrasts of wet and dry areas within their political boundaries as a kind of distributional error that they aimed to solve such that not a drop of water would reach the sea.[84]

Colossal labyrinths of concrete and canals have not alleviated pressures on groundwater in either locale. In some of California's prime agricultural districts, groundwater tables are falling by an average of two to four feet per year.[85] The land is literally sinking as aquifers compact, in some places by as much as 10.6 inches

(270 millimeters) in a single year.[86] The state's landmark Sustainable Groundwater Management Act of 2014 has yet to be fully implemented, and its long-term effects are uncertain given the seemingly intractable nature of groundwater management in arid, agriculturally dependent economies.[87] In Spain, despite policy mechanisms to restrict groundwater withdrawals since 1984, irrigated agriculture continues to deplete aquifers at an accelerating pace, as politicians are reluctant to oppose profitable agro-industries and illegal wells are widely acknowledged to be widespread.[88] Growers in both California and Spain have been leaders in the adoption of high-efficiency irrigation systems, and both locations clearly demonstrate a "rebound effect" in which these technologies have heightened water withdrawals by amplifying intensification (see chapter 2) and enabling expansion into previously nonirrigable districts.[89] Climatologists anticipate water stress increasing as both regions become hotter, droughts last longer, snowpack in headwaters becomes less reliable, and precipitation becomes more variable. In short, these two fruit and vegetable baskets that blossomed during the late twentieth century both appear exceedingly fragile.

The uncanny parallels between intensive agriculture across these two regions are not entirely explained by climate. The links between these rural landscapes are also distinctly geopolitical and institutional. Back in 1953, when European countries in the wake of the Second World War were still isolating Spain to punish Franco's alignment with the Axis powers, the United States made an abrupt change of course. The Cold War with the Soviet Union had become US foreign policy's top priority, and Spain's strategic location superseded ideological opposition to Franco's dictatorial regime. US President Dwight Eisenhower signed a series of agreements with Franco known collectively as the Pact of Madrid, which allowed the United States to construct several military bases on Spanish soil in exchange for economic and military aid. One component of this economic aid came in the form of helping to reproduce the agricultural extensions service model developed by US land grant universities.

Franco's agricultural minister was invited to visit the United States for a tour and left so impressed that upon his return he immediately issued an order to create Spain's own extension service and sent two agents to study the institution for six months.[90] A report in 1960 indicated that 45 percent of the Spanish extension program's initial budget came from the United States, ratcheting down to 30 percent over the first decade.[91] Even before this national partnership took hold, Franco had been deeply inspired by the use of irrigation to encourage agricultural settlement in the western United States, using it as a model for his own technocratic scheme to establish irrigated colonies to settle rural lands.[92] Taking US agricultural research and outreach institutions as a template was a logical next step. While the farmer-facing Servicio de Extensión Agraria was largely disassembled around the time Spain entered the European Union, its core research institutions remain. Given the shared interest in fruit and vegetable crops, scientific exchange between

Spain and California is robust. It is all the more striking then that almonds, which had become a paragon of California's ability to harness science for agro-industry during the late twentieth century, remained largely outside Spain's scientific scope of interest. Despite water infrastructure and agronomic institutions modeled on California's, and intensive horticultural agribusiness sectors that parallel each other in many ways, almonds in Spain have persisted at the social and ecological margins. In an agricultural economy accelerating intensification on so many fronts, almonds have appeared to be spectacularly immune.

As I became progressively more fixated on the contrast between almond ecologies in California and Spain and began digging into the archives, I never imagined I would find that someone else had actually beaten me to it. In 1921, the recently formed California Almond Growers Exchange published a glossy booklet distributed to US congresspersons titled *Shall the American Almond Industry Perish?*. In it they pleaded for tariffs to be imposed on almond imports, overwhelmingly coming from Spain. The photo-rich layout of each spread depicts the "pauper labor" and "neglect" of Spanish production on the left page contrasted with the "intelligent care" of California's "high-class land" on the right.[93] Page by page, a case is laid out. "Minimum efficiency" donkey-pulled carts in Spain are contrasted with "the inventive genius" of gasoline-powered, tractor-mounted carts in California. Nut-shelling by elder Spanish women using "home-made" equipment with materials "donated by nature" are juxtaposed against California's brand new, brick, almond-processing factory with "scrupulous sanitary conditions." On one page, a pair of Spanish growers in weathered clothes, labeled as "Descendants of the Moors," are depicted squinting uncomfortably into the sunlight framed by the doorway of a rustic earthen home, while the opposing page portrays a team of sharply dressed young men using "scientific cultivation" to spray almond trees on the University of California's experimental farm. "Perfect almonds are a product of skill," the caption reads, "nothing is left to chance." The text is suffused with a condescending attitude toward Spain's peasant producers. Its opening page concludes, "God forbid we stay in business long enough to be forced to adopt the living conditions of the bulk of our competitors." As it turned out, comparison between California and Spain was not merely an academic endeavor that I had externally imposed. In fact, it was a significant feature of California's first almond overproduction crisis, a profitability crunch resulting from the domino effects of surplus settler mining-industry capital plowed into agriculture. Once almond harvests swelled beyond what local markets could absorb, industry leaders turned to comparisons with Spain as the centerpiece of their appeal to policymakers.

The booklet makes clear how deeply the state is implicated in California's agrarian capitalist developments. From the very first page, the authors note that their primary concern is the threat to "landed investments," which would become "liabilities rather than assets" without state action. There is little bucolic romanticism. The stakes are articulated in terms of balance sheets rather than rural communities.

And despite the entrepreneurial dynamism proclaimed in their materials, CAGE spells out just how much the California industry's success is predicated on state intervention. "Three hundred thousand dollars have been invested" in a modern almond processing plant, the text reads, "on the assumption that the Federal Government would not permit the extinction of an essential American food-producing industry." Due to unrestricted imports, the description continues as it pins the blame on regulators: "THE PLANT IS NOT NOW OPERATING [*sic*]." California growers and processors were not the only ones requesting state intervention in almond markets to salvage their business, but they claimed that major investment decisions had been made on the assumption that the state would protect their financial interests. Given the proper tariffs in place, they optimistically projected, "The business is capable of being extended almost indefinitely."

The California growers' plea was successful, despite opposition by confectioners and merchants sourcing cheaper imported almonds. A literal side-by-side comparison, which denigrated peasant production techniques and effectively racialized Spanish producers, was an influential component in garnering state support to (temporarily) alleviate overproduction. The comparative framing insisting that almond landscapes be treated as objects of industrial optimization, rather than free gifts of nature, was not simply a matter of idolizing scientific rationale. It candidly illustrated the ways in which industrial management is necessary for almonds to continue as a lucrative vehicle for investment. The low cost and minimal effort of Spain's household production was a thorn in the capitalists' side that only the state could remove. "Four thousand years of 'go as you please' Almond production was checked with American Horticultural skill and effort were brought to bear [*sic*]." Comparison was a rhetorical trick of the trade with very material consequences.

DIFFRACTION AS COMPARATIVE METHOD

Comparison has been a theoretical challenge for post-positivist social scientists. Much like the comparative booklet by California almond industry leaders, sociological analysis has a history of deploying comparison to admonish and discipline those who exist outside of Eurocentric notions of modernity.[94] The plotting of nations along a hierarchy of "development" endures in many public arenas. Even those challenging such colonial legacies may fall into the same trap, such as scholars using world-systems theory that rests on a distinction between the core and the periphery,[95] effectively reproducing the "West and the rest" narrative.[96] Responding to these concerns, scholars across the social sciences have ventured to develop comparative approaches that can dispense with essentialist categories, ground themselves in historical specificities, illuminate interconnections, and foreground the active coproduction of social worlds. Sociologist Phillip McMichael proposes "incorporated comparison" as a method that rejects the idea that researchers are tracing the trajectory of preestablished phenomena like globalization or capitalist development or that units of analysis are naturally given or easily comparable. Instead, the task is to construct an understanding of capitalist political economic processes through nondirectional, fluid, contingent historical relations.[97] Geographer Gillian Hart likewise takes issue with elements of the Marxian tradition that presume cases are variants of ideal types or evidence of a generalized process, particularly those that diminish the active role of race, gender, and other forms of social difference. Drawing on Henri Lefebvre and Doreen Massey, she offers "relational comparison" as a way to advance a comparative method that centers the active social construction of space and place.[98]

At a time when anthropologists were actively distancing themselves from comparison, Marilyn Strathern became its champion, arguing that all ethnographic work was rooted in implied comparisons that should be made explicit rather than concealed.[99] Her proposed method justifies some simplification, and even playfulness, for the purpose of being provocative and challenging common sense.[100] Like Haraway's "situated knowledges,"[101] Strathern argues that admitting the partial nature of connections offers analytical strength rather than weakness. A parallel thread of comparative theorizing, though it may not be labeled as such, examines different "cases" as internal to a given ontological object. The ontological turn is a strand of scholarship that details how the nature of reality itself can be multiple, rather than merely interpreted through varying perspectives. Anthropologist and philosopher Annemarie Mol lays out such a method in her book *The Body Multiple*, which encourages researchers to analyze how distinct understandings of reality hang together and to show the work that is required for their coordination.[102] These approaches flip positivist comparison on its head. Rather than reify preordained categories or objects, comparison serves to denaturalize established common sense and suggests this type of disorientation is analytically illuminating.

My approach to comparison has grown out of a deep appreciation for these reflexive, creative, textured ways of thinking two or more things at once. My thinking has also been profoundly shaped by the theory of agential realism that feminist science and technology scholar Karen Barad offers and how it might serve comparative analysis. In her exploration of quantum physics, *Meeting the Universe Halfway*, Barad argues that existence as we know it is not an assemblage of interacting parts but rather that all objects are actively produced in relation to one another by making "cuts" while they remain fundamentally inseparable.[103] These cuts are not merely social constructions projected from the minds of scientists or others onto a passive world, they are actively coproduced with the material properties of biological and physical phenomena, including those of human bodies. Taking it a step further, she insists that matter and meaning are always inherently intertwined. I agree with these claims and hope to operationalize them in the chapters that follow.

In reading Barad's work, I was particularly struck by the term *diffraction*, a visual illustration of how the wavelike properties of light bend and reshape one another as they interact. Donna Haraway had previously offered the visual metaphor to describe how "interference patterns can make a difference in how meanings are made and lived."[104] Rather than reflecting a phenomenon, diffraction patterns record a passage or process.[105] Barad expands this type of exploration by proposing diffraction as a methodological approach "of reading insights through one another in attending to and responding to the details and specificities of relations of difference and how they matter."[106] This sounded like an excellent description of what I set out to accomplish in my own research. Diffraction was both something I was witnessing, as I traced the various ways the histories of California and Spain rippled out in relation to one another, and something I was actively doing, as I forced together cases that most of my interlocutors insisted were irreconcilably distinct.

Barad's work, for me, adds two important elements to the nonteleological, nonessentialist, relational comparative approach that Hart and colleagues have offered. One is that it brings the uneven contours of the comparative research process more candidly into view. When I began this project, I felt self-conscious about the way that everything I was learning about Spain was filtered through my own experiences in California, and then later on, how my reading of California's almond landscapes became irreversibly changed by what I had learned and experienced in Spain. As I began writing, I worried that the book would appear lopsided, and I was preoccupied by how to treat each case similarly. But there is no point in pretending that I chose this comparison from a neutral detached position or that its contents are meant to speak equally to both geographies in the same ways. This book is not only an exercise in carefully tracing the relational, contingent practices and processes of agricultural knowledge across two interrelated cases, it is very explicitly about using each to bend our perceptions of the other. The phenomena I describe are not only relational by default, my analysis also actively generates new relational interference patterns, reading cases not side by side but *through* one another.

If it appears at times that the Spanish case is being instrumentalized to understand damaging practices in California or that the California case is being instrumentalized to recognize the virtues of rainfed agriculture in Spain, it's because it is. While I maintain a commitment to empirical depth throughout my work, the pursuit of "balance" ultimately came to feel disingenuous. I would also be failing in my duties not to acknowledge the influence of my positionality as a researcher based in the United States and a member of a California-based institution, drawn toward this research by the rapid expansion of almond agriculture in California. Diffraction helps me to explain that taking a comparative approach does not mean I must provide an equivalent, meticulous accounting of perfectly parallel phenomena in each case. Instead, the objects of my analysis are the interference patterns generated through the act of comparison itself, embracing their lumpy, partial, situated, and political nature.

METHODOLOGY AND PROCESS

The sequence of this research was shaped a bit like a boomerang. I began exploring the political ecological contours of the California almond industry in the fall of 2015 during the second year of my doctoral program at the University of California Santa Cruz. I was fascinated by the flurry of media attention to the almond industry and wanted to understand how it had come to epitomize the state's water problems despite the fact that, as industry representatives repeated to anyone who would listen, almonds do not draw more water per acre than similar tree crops like plums or peaches and are not the most water intensive crop (alfalfa takes that prize). Rather than dismiss this media attention as a product of ignorance, I found this incongruity even more compelling. In analyzing newspaper articles spanning two years, I found that almonds attracted attention not simply because of how much water they used but because they exemplified discomfort with the enormous power markets had come to wield over a public good.[107]

Upon learning that Spain's rainfed production had dominated the global industry just a few decades prior, I became enthralled by the way in which a single crop's agroecosystem could be understood in radically different ways: the poster crop of unquenchable thirst in one place and of unrivaled thrift in another. In December of 2015, I attended The Almond Conference hosted by the Almond Board of California in Sacramento, which would become the first of five for me. The annual gathering proved an invaluable opportunity to speak informally with growers, vendors, and agronomists while witnessing knowledge politics in action through educational sessions on irrigation, beekeeping, marketing, policy, and sustainability programs. In the summer of 2016 I visited UC Davis Library Archives and Special Collections and dipped a toe into full semistructured interviews with a few almond growers in the Sacramento Valley.

In September of 2017, I traveled to Spain and began making research contacts while compiling archival materials at the Ministry of Agriculture, Fisheries and Food in Madrid. Over the course of ten months, I spent time in eight different comarcas, roughly equivalent to provinces, with particular significance for the almond industry past and present. In Alicante (Valencia) and Mallorca (Balearic Islands), I witnessed the way diversified rainfed production systems, which once formed the core of Spain's almond production, persisted as they were progressively overpowered by coastal tourism.[108] In Granada (in eastern Andalusia) and Murcia, I engaged with rainfed production in the high plains, seeing the legacy of twentieth-century production systems as well as new plantings inspired by the sky high prices of 2014. There, in addition to interviews, I joined pruning classes and other events for farmers new to almond production. In Córdoba (western Andalucia) and Lerida (Catalonia), I saw how prime irrigated land was being converted to almonds at a dizzying pace, drawing from groundwater, existing surface water infrastructures, and the recently completed Segarra-Garrigues Canal. In Huesca (Aragón) and Zaragoza (Aragón), I saw a combination of rainfed plantings tucked among the foothills of the Pyrenees alongside new small-scale irrigated orchards. I wished I could have spent time in Albacete (Castilla–La Mancha), a flat interior region with almost no prior almond growing history that has become, since the time I was doing fieldwork, a major site of the almond boom. I tried to identify beekeepers who worked as pollination service providers for almond operations, but all the intensive orchards I visited used cardboard boxes of bumble bees, which are used for a single season and then thrown away (see chapter 3). Spanish agriculture is incredibly heterogeneous topographically, economically, culturally, and otherwise, and I clearly packed an enormous range of field sites into a relatively short time. Such limited time at each site carries significant limitations, particularly regarding the nuances and historical depth of political-economic relations in play. I hope that what I sacrificed in depth I gained in breadth.

My method for identifying farmers varied at each site. My goal was to capture the broadest range of almond farming experiences in a given area, including plantings of varying ages, sizes, irrigation styles, and levels of diversification along with the demographic and socioeconomic characteristics of those who cared for them. Most often, I would contact a local almond producer cooperative, interview one or more agronomists there, and ask to be connected with farmers across a broad spectrum of circumstances. Whenever possible, I would attend workshops and events dedicated to almond production and network with farmers who were participating. Many whom I interviewed were willing to recommend friends, allowing the sample to "snowball." In three different locations, families who rented me a room in their home and were intrigued by my work offered to connect me with almond producers they knew. In the case of corporate almond plantings, which had websites for investors, I would contact the company directly. I always asked my interlocutors if they would be willing to walk or drive around their property, treating the orchard landscape not simply as a backdrop but as an active participant in the interview process that raised new questions.[109] Seventy-six farmers participated in this portion of the study, nineteen of whom were also working as agronomists or administrators at local cooperatives.

Central to my methodology were interviews with researchers at agricultural science institutions with a significant history or active role in almond research. These included CEBAS in Murcia, CITA in Zaragoza, IRTA in Catalonia, IFAPA in Córdoba (western Andalusia), and La Estación Experimental del Zaidín in Granada (eastern Andalusia). I tracked down several prominent almond researchers in their retirement. In addition to my core focus on almond farmers and researchers, I made trips to locations with special importance for the almond industry. I accessed the archives of the Colegiata de Santa María la Mayor in

Alquézar (Huesca) where a priest in the early twentieth century conducted early breeding experiments (see chapter 1). I spent time in Jijona (Valencia), the center of production for Spain's famed almond confection turrón, where manufacturers previously touting the distinctive qualities of Spanish almonds had recently shifted to importing nuts from California.[110] In Madrid I visited the offices of the almond export industry group Almendrave. Sixteen researchers and twenty-six additional industry members, such as processors, equipment providers, and turrón industry professionals took part in the Spanish portion of the project. My interviews took place in almond orchards of all kinds, as well as vehicles, coffee shops, homes, offices, experimental plots, and public squares.

When I returned to California in June of 2018, I saw the densely packed almond orchards of the Central Valley with fresh eyes. The notion that they "required" fifty-six inches of water or two colonies of bees per acre felt surreal with the dust of rainfed fields still on my jeans. Over the course of eight months, I worked to interview growers and industry professionals across two main geographical regions: the relatively wetter northern and central counties where almonds had been concentrated during the early twentieth century (Colusa, Glenn, Butte, Yolo, Tehama, Sutter, Stanislaus, Merced, San Joaquin) and the more arid southern counties where almond orchards were rapidly expanding and intensification methods were being continuously perfected (Kern, Fresno, Madera, Tulare, Kings). Given the negative press surrounding the almond industry during the state's historic drought, it was understandably difficult to find California almond growers willing to be interviewed by anyone, let alone by a social scientist in an environmental studies department coming from an institution with a strong leftist reputation. This posed a significant challenge. The staff at the Almond Board of California provided recommendations to speak with farmers who had participated in their almond ambassador's program, but of course I wanted to conduct many more interviews that would be less directly shaped by industry talking points. My most successful recruitment method was cold-calling farms from a publicly available business information database called Buzzfile. Out of one hundred forty-six contacted farms, twenty-nine chose to participate in an interview. Ten of these interviews represented the north-central region, and the remaining nineteen represented the south, which accounted for 56 percent of production in 2022. I also visited a region that was California's leader in rainfed almond production during the early twentieth century, Paso Robles, and spoke with two landowners with remnant orchards. Grower interviews took place in a wide variety of venues, including coffee shops, homes, real estate offices (many California almond growers, I learned, also work as land valuation consultants), over Zoom, and even once in a private airplane hangar.

Beekeepers who provide pollination services for almond growers offered enormous insight not only into their own work but also into the almond industry as a whole and its progressive intensification over the years. The twenty beekeepers

I interviewed all ran full-time commercial operations ranging in scale from seven hundred fifty to twenty-four thousand hives, and one who preferred not to disclose but explained his outfit was one of the biggest in the business. All were California-based save for two who had recently begun trucking their bees from Georgia and Florida for almond pollination season. Roughly half of these interviews took place in person at beekeepers' homes or offices while the others took place over the phone, given the wide geographic range they travel over the course of the year.

California has a single academic institution producing the lion's share of almond-related research, the University of California Davis, alongside a network of agricultural extension offices throughout the state. Eight researchers and extensionists with high-profile work in almond breeding, irrigation, pollination, and nutrient management, including some in retirement, graciously accepted my interview invitations. When possible, I also attended almond grower education events offered by the extension service. The Almond Board of California, well-versed in public relations, agreed to interviews with their executive team, including the CEO, chief science officer, director of sustainability and environment, and vice president of global marketing. To better understand the growth and strategies for market development, I spoke with two marketing professionals at Blue Diamond, the largest almond grower cooperative. I also interviewed a water justice community organizer and two almond processors as tangents along my travels through the Central Valley.

The empirical material informing the analysis in this book is wide-ranging, and my methods in Spain and California could never have been perfect mirror images of one another. Instead, I offer a comparative study that embraces the lumpy, partial, systematic, and yet serendipitous reality of qualitative research.

THE ALMOND PARADOX IN FOUR ACTS

This book teases out the politics of what plants need by tracing the diffractive trajectories of four topics: breeding, water use, pollinator relations, and place in California and Spain. While each chapter stands alone, transitional moments in each chapter occur sequentially, layering each historical narrative on those in the preceding chapters. I begin with the breeding of foundational varieties in the late nineteenth century, then the codification of crop water requirements in the mid-twentieth century, followed by the transformation of beekeeping into a pollination services business in the early twenty-first century, and finally the shifting of geographies of production with an emphasis on the boom of the last ten years. The discussion of spatial shifts also turns toward alternatives to intensification by illustrating the current struggles among Spain's rainfed farmers and hopes for enrolling almonds in rural revitalization.

Chapter 1, "Matter: Meaning-Making in a Nutshell," takes up the question of plant breeding in each place as a means of revealing how values are baked into the very molecules that make up the almond tree. I begin by laying out the historical foundation of contemporary almond cultivation in the late nineteenth-century colonial relations of California and the rigid hierarchies and land struggles of rural Spain. Using the writings of the most influential plant breeder in each place—a California capitalist using orchards to promote white supremacy and a Spanish cleric serving the poor—I show how the preference for soft-shell almonds in California and hard-shell almonds in Spain would carry lasting impacts for pesticide use over a century later. In doing so, I bring the concept of material semiotics from feminist science studies into conversation with the political economy of agriculture. Here coloniality and social marginalization are quite literally embedded in a nutshell. What matters to those pursuing plant knowledge can be read in the materiality of the plant itself.

According to current agronomic science, almonds in California's top producing regions "require" fifty-six inches of water. Chapter 2, "Flow: Knowing Plant-Water Relations," unpacks the measurement of "crop water requirements" via evapotranspiration (ET), revealing how a constellation of capital, technologies, state institutions, and farm ownership structures shape the knowledge of plant-water relations. Using archives and interviews, I show how microirrigation techniques and university research operated in tandem with the rise of corporate and absentee farm ownership to rework and rapidly disseminate increasingly intensive irrigation standards. As a casualty of such knowledge practices, agronomic calculations render rainfed almond cultivation practices illegible or even illegitimate. While California's intensification took decades and immense public resources, amid the boom, codified "crop water requirements" were transported to Spain nearly instantaneously. Actual water use in Spain is a closely guarded secret, so I use yield data comparisons to demonstrate that new irrigated orchards are replicating California's extractivism. Despite the hegemony of "requirements" based on ET, rainfed almond farmers in Spain continue to cultivate more ecologically suited knowledge practices of plant-water relations using a range of water conservation techniques. This chapter emphasizes not only the politics of seemingly objective irrigation calculations but also the propensity for such knowledge forms to jump scales as capital rushes in.

California's almonds orchards are now dependent on the largest managed pollination event in history each February, drawing nearly every commercial hive in the United States into their orbit. These hardworking honeybees are dying at rising rates each season, stressed by the four Ps: pesticides, parasites, pathogens, and poor nutrition. Meanwhile, relocating across the country for pollination services has now surpassed honey production as the economic engine of the beekeeping industry. But what does it mean for a crop to be pollinator dependent, and how did this industrial symbiosis come about? In Spain, as in California historically, growers did not see a need for bees or kept a few hives themselves. In chapter 3, "Symbiosis: Producing Pollinator Dependence," I trace the surprising history of pollination services for almonds, once driven by beekeeper's desires to intensify their own business and only later encouraged by orchard agronomists to boost almond yields. Drawing on interviews with beekeepers, I describe the consequences of a deepening dependency on almond pollination, including feeding bees corn syrup, soy patties, and regular doses of agrichemicals to meet grower expectations. I then turn to Spain where growers say they simply don't need bees given the agroecological context of marginal landscapes. Pollination has been a central concern for breeders, however, and a growing minority of farmers also benefit from self-compatible almond varieties that do not require cross-pollination with another variety to fruit. In contrast to irrigation, here the almond boom pushes California growers to adopt technologies from Spain, as they seek to lower pollination costs. However, the scale of industrial production and market conditions quickly reign in enthusiasm for self-compatibility. Thus I show how pollinator dependence is no mere botanical description and can't be remedied by a technical solution. Claims about plant-pollinator relations reflect a complex political economic dance that transcends physiological explanations.

Whereas previous chapters deal with knowledge surrounding what, how, and with whom almonds are produced, chapter 4, "Space: Creeping Toward Precarity," examines notions of where almonds should or should not grow. The California almond industry has been accused of growing in the wrong place, effectively a desert, yet I show how the shift toward more arid conditions makes perfect agronomic sense. Aridity creates factory-like conditions where inputs can be carefully controlled without the vagaries of weather. In Spain, rainfed growers too are faced with assertions by buyers that they are growing in the wrong place, at the unmanageable margins, and a shift to intensified irrigated valleys is lauded as bringing efficiency. This shift not only exacerbates such burdens but also poses a problem of ecological instability for the landscapes where almonds are falling into abandonment. Thinking across these two cases, I illustrate how profitability draws production into spatial configurations that deepen their precarity: in California toward aridity and in Spain toward overburdened aquifers. The location of almonds in the seemingly wrong places is not an accident of history, it is the production of clear agronomic logic. This, I argue, necessitates a reconsideration of how to root agricultural knowledge in the specificities of a given place.

This book does not delve into the intricacies of almond consumption. There are three main reasons for this. One is that if a reader comes away from this text with the conclusion that they should eat less almonds, then I will have failed. I am not writing this book to talk you or anyone else out of their almond milk latte. Enjoy it. I want to raise questions about how industrial agriculture's environmental precarity is baked into knowledge production processes. This would be true about any crop undergoing capitalist pressures to churn out profits without regard for socio-ecological peril. Almonds happen to be an illustrative case. The second reason not to spill much ink on consumers' food choices is because the California Almond

Board has spent a mind-boggling amount of marketing dollars shaping them. Nearly all nutrition research specifically dedicated to almonds has been funded by the industry in some way.[111] While foods rise and fall in popularity for a variety of reasons, I view the recent surge in appetite for almonds largely as an outgrowth of an ever-looming overproduction crisis. And when consumers move away from almonds, or markets are unable to absorb mounting surpluses, the next boom crop will simply take its place.[112] Finally, consumer-centered food system critiques tend to offer unsatisfying neoliberal solutions like buying ethically sourced products.[113] Certainly purchasing decisions make a difference, but the idea of "voting with your dollar" narrows the scope of the political imagination and gives more votes to those with more dollars. Instead, I hope to demonstrate that recovering from extractive agricultural practices would be better served by structural and institutional changes. Reigning in the role of science in the treadmill of overproduction, valorizing "marginal" agroecosystems, and rooting agronomic knowledge production in the specificities of place are directions drawn from this analysis that offer a great deal of hope. To build just and thriving agri-food futures, we must change more than what we eat. We must change what and how we *know*.

1

Matter

Meaning-Making in a Nutshell

The most fundamental form of plant knowledge is an investigation and understanding of the plant body. How it behaves. What characteristics it exhibits. In studying these elements, one can pick favorites among the multitude of natural variations and choose to propagate those with the most desirable characteristics. This process of close observation, discernment, and choice by individual farmers is as old as agriculture. Stabilizing and reproducing a variety with a designated name, however, is a relatively recent phenomenon. When farmers grow primarily for their own consumption or local markets, produce their own seeds, and rely on relatively little processing machinery, the consistency of reproducing a single named type is largely unnecessary. As the late nineteenth century shifted many farming operations toward global markets and mechanization, however, isolating and reproducing plants with a set of known characteristics became the linchpin of successful farming. Before the agricultural sciences launched into methods for determining how much water, fertilizer, or pollinator activity a crop may "need," those tasked with improving agricultural production focused on what a plant *needed to be*. This meant selecting the best cultivar, a portmanteau of cultivated variety.[1] Plant breeding serves as the foundation on which all other agronomic inquiries build.

Knowledge and selection of agricultural plant matter is inescapably political. Breeding is bound up with ideas about how land should or could be transformed, who will be engaging with it under what circumstances, and what distribution of impacts the resulting agroecosystem will entail. As such, plant breeding has been a central concern for scholars of agrarian change. The popularization of hybrid corn in the 1920s handed private seed companies substantial power over farmers and public institutions while positioning agribusiness as a source of scientific

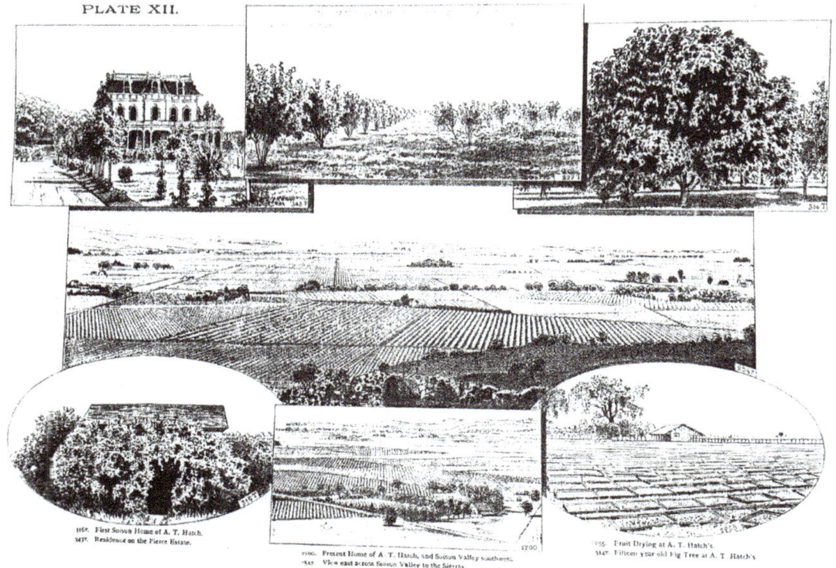

FIGURE 1. Photographic plate from the profile of A. T. Hatch, renowned Nonpareil almond breeder and orchard capitalist, for the Vacaville chapter of the promotional book *California Illustrated*, by Edward Wickson (California View Publishing Company, 1888). Courtesy of the Vacaville Museum.

FIGURE 2. Remnants of the experimental almond orchard used by Father Rafael Ayerbe Castillo to breed the Desmayo almond, located behind the Colegiata de Santa María la Mayor in Alquézar, Spain. Photo by author.

authority.[2] Genetic engineering of crops effectively rewrote intellectual property law and exacerbated the commodification of seed.[3] More recent scholarship has centered the political implications of specific breeding priorities. Dwarf apple varieties reflect the precarity of accelerating market cycles.[4] The pursuit of large, vibrant, and high-yielding strawberry varieties has exacerbated growers' dependence on poisonous soil fumigation.[5] The relations among people, plants, and places carved out by plant breeding exemplify what many geographers have come to call hybridity,[6] a fusion of human intention and action with the autonomous dynamism of the biophysical materials in play.

Crop breeding is frequently described as improvement, indicating a singular destiny, but what counts as a good outcome is highly specific to the circumstances of a given place and time. Any given variety presents tensions and trade-offs. Actors along the supply chain, from farmers to processors to retailers to eaters, have differing priorities. The plant's physiology, genetic characteristics, and life cycle shape possibilities and often generate unexpected results. Generations of breeding layer interwoven histories of desire and happenstance upon one another. In this way, plant varieties can be read almost like texts.[7] The characteristics on display in a given crop—fruits of a certain color or shape, leaves attaining a given form, roots exploring the soil with a distinctive pattern—reveal a complex fusion of botany and political economy. What matters for plant breeders, and the complex array of interests they seek to align, is baked into the very plant matter itself.

Plant varieties carry the legacies of the past into the present. In fact, for the almond and other related woody tree species, the plants alive today are largely clones of their progenitors. They have been reproduced from cuttings that contain the exact same genetic material as their parent plant. An almond variety selected a century ago may encounter radically different social and agroecological conditions today, yet the values that guided its initial development live on in its very cells. The material qualities it possesses embody meanings.[8] Once a cultivar is popularized, it becomes a living manifestation of interactions among atmosphere, microbes, plants, breeders, nurseries, growers, and buyers built on centuries of both intentional and unintentional selection. As with other technologies, plant varieties can generate path dependencies and incur unanticipated consequences. While new types of significance may be attributed to a given variety over time, the legacies of its origins always remain etched in its DNA, like an unchangeable ledger.

As I was sifting through the archives of almond research from the late nineteenth and early twentieth centuries, I was struck by the politically charged nature of the breeding activities in both California and Spain. In each place, a single variety stands out as the most enduring, in many ways setting the course for the industry as a whole. And behind each of these famed varieties was an individual who invested their energy in improving almond cultivation to serve a specific ideological agenda. The Nonpareil in California was selected by a wealthy orchard capitalist who saw the profitability of almonds as essential to a racial project for recruiting Anglo settlers to replace a predominantly Chinese workforce. By contrast, the Desmayo Largueta in Spain was developed by a priest who saw almond trees as a stabilizing force for the rural poor, enabling them to endure harsh inequities without challenging the status quo. He hoped that an investment in private property would dampen the appeal of communist ideals. The motives of these two men are manifest in the form that each of their prized varieties took, in California a thin-shell nut to maximize profit potential and in Spain a hard-shell nut formed from a drooping flower to provide frost protection. These forms carry cascading effects a century later. California's Nonpareil sets a course of deepening pesticide dependency, whereas Spain's heterogeneity of frost-protective, hard-shell varieties privilege resilience over productivity.

This chapter's tale of two breeders takes a diffractive approach, using the case of each to expose and amplify the significance of the other. Each historical narrative on its own tells a compelling story about situated plant knowledges. But it is only by seeing one historical narrative from the perspective of the other, seeing Spain through the eyes of California and vice versa, that something as simple as a nutshell can so clearly illustrate a bundling of matter and meanings, path dependencies and paths not taken. This relational approach is befitting of plant breeding, where a given cultivar can never exist in a vacuum. It can be evaluated only by how it stacks up against its peers. Rather than a comprehensive history of almond breeding,[9] my aim here is to focus on two game-changing moments where new forms of almond physiology embedded with ideals enabled distinctive relationships to coalesce around the orchard. In doing so, I aim to illustrate how plant matter itself is a form of knowledge politics. Relationships between land, labor, capital, the state, insects, fungi, water, and more are woven into the fibers of nutshells and the forms of flowers. Matter embodies meanings and shapes social legacies, with and beyond intentionality. Agricultural knowledge politics is sedimented in the biophysical substance of the farm itself, often with unplanned outcomes long after charismatic protagonists move on. Almond varieties and their ongoing legacies show us the material afterlives of ideas and their political economic underpinnings.

This argument feeds into a broader conversation about how settler colonialism operates through the refashioning of land and lifeforms. In the case of almond breeding around the turn of the twentieth century, we see two unstable edges of a colonial configuration. When California's famed Nonpareil came into being, California was on the growth edge of the US settler colonial project. It was a recently established state violently appropriating land from Indigenous communities,[10] using legal antics to uproot Mexican ranchers,[11] and recruiting settlers from farther east to bolster the population of white citizens. When Spain's Desmayo Largueta came into being, the country had recently relinquished the last of its colonial territories to the United States in a catastrophic war and was turning inward to its rural hinterlands in hopes of reimagining its sources of identity and wealth.[12] The prioritization of resilience in Spain's almond selections may seem ecologically savvy and altruistic, but it can also be read as deeply conservative and paternalistic. This is not a story of guilt versus innocence. It is a pairing that illustrates the many ways that colonialism, whether at its unstable expansionist frontiers or its internal reckoning with reluctant retreat, seeps into plant knowledge practices and the very plants themselves.

CALIFORNIA'S NONPAREIL:
UNEQUALED PROFITABILITY

The Nonpareil almond has come to represent the industry standard. Smooth, light-colored, and elongated, it is both picture-perfect and highly productive, the California almond industry's champion for nearly 150 years. It would likely make up even more of the volume and acreage than it does were it not for the fact that it must be planted in alternating rows with a distinct variety to be successfully pollinated. It is difficult to overstate the importance of the Nonpareil. While other varieties have their distinct niches, the Nonpareil remains the benchmark against which all others are judged. Nearly every orchard management practice in California, from the design of the orchard to harvest scheduling, along with processing machinery and marketing campaigns, revolves around the attributes of this one, quite literally unequaled, cultivar.

Today's Nonpareil almond trees are clones of a serendipitous sprout that emerged in the fields of a silver miner turned agribusinessman in 1879.[13] Augustus Timothy Hatch had accumulated a fortune from the Reese River silver mine in Nevada before investing his newfound wealth in a ranch in Suisun, California, a small settlement named for the indigenous Suisun people, who were violently forced to relinquish their sovereignty just decades before. California had been claimed by the Spanish since 1769, presided over by Mexico beginning in 1821, and seized as a spoil of war by the United States in 1848, but it was in Hatch's orcharding heyday (the 1870s and 1880s) when Anglo settlement of this vast landscape became a focused political project and killings of native peoples reached their peak.[14] It was also a moment of intensifying anti-Chinese bigotry as white laborers, along with political leaders, viewed immigrants from East Asia as a threat to white prosperity and racial dominance of the territory.[15] This historical context was not incidental to Hatch's almond cultivation activities; in fact it served as a core logic.

Demonstrating dazzling levels of agricultural profitability was considered a crucial tactic for enticing white settlers to make California home. Hatch helped organize and served as president of California's State Board of Trade in 1887, a voluntary organization formally incorporated three years later. The Board aimed "to make California better known and thus to attract desirable immigration"[16] and was largely tasked with "supplanting Chinese labor with white labor."[17] The waves of migration to California following the state's 1849 Gold Rush included many arrivals from China who then faced repression, segregation, and racial violence, eventually resulting in the 1882 Chinese Exclusion Act. White working-class men blamed Chinese workers for their economic woes, and throughout the state Chinese communities were repeatedly expelled, denied services, attacked, and subjected to arson.[18] As theorists of racial capitalism have illustrated,[19] the ability to accumulate capital is predicated on producing and exploiting hierarchies of difference. Hatch was quite charismatic in using his business success to advance a project of white dominance, but his work with the California State Board of Trade was entirely unexceptional.

While Hatch's contemporaries earned reliable returns from wheat and cattle rearing, he was determined to push profits higher by focusing on fruit trees.[20] When his initial crop of wine grapes did not prove sufficiently profitable, he turned to almonds.[21] Instead of using the most popular variety sold in nurseries of the time, the Languedoc, brought over from France, he planted the almonds of the Languedoc directly into the soil to see if he might get an even better crop. Direct seeding was uncommon because it can produce wildly variable results. An almond tree must be reproduced vegetatively to retain its desirable traits and then grafted onto a rootstock that suits local soil conditions. The tree's highly diverse offspring are an evolutionary strategy common to plants from harsh environments where genetic diversity offers greater advantages. When Hatch found one of the Languedoc seedlings satisfactory, he set out to graft it onto a set of bitter almond rootstocks he had seeded.[22] Hatch had overestimated the number of rootstocks he would need and wound up with two hundred extras, which he allowed to mature out of curiosity. One of these spare rootstock seedlings, never originally intended to yield nuts, produced the now ubiquitous Nonpareil. Hatch was never trained in agricultural sciences, but he aimed to spot in the earth's dynamic processes a financial opportunity and then spread the message in order to root California's settler colonial society.

Paper-Shells, Profits, and Racial Projects

The Nonpareil[23] shows many attractive features that might explain its nearly 150-year reign. It is consistently productive, adaptable, visually appealing, and lacking in significant flaws. But during Hatch's time, it offered a major innovation that set it apart from anything on the market: an extraordinarily thin shell. So thin in fact that it earned the description "paper shell." This ultrathin shell meant that when a grower went to sell almonds, they could fetch a much higher price per pound based on the fact that a greater percentage of each bag's weight was the actual edible nut relative to the shell. California almond growers quickly began to recognize that to reap plentiful profits, their varieties needed to have a "clean, thin, soft shell" and "a smooth, bright, plump kernel."[24] Whereas the more common hard-shell varieties sold for 12 cents per pound in 1891, softer- shelled varieties could fetch 16 cents and paper-shell varieties a whopping 22.5 cents per pound.[25] The standard-bearer of the time, the Languedoc, contained only 30 percent nutmeat by weight, but the

Nonpareil boasted 60 percent. Payment for Nonpareils was, accordingly, nearly double. Hatch stated frankly, "I regard the [Nonpareil] seedlings as better than the Languedocs, for the reason that they bring more money."[26] Another orchard capitalist underscored the financial boon of Hatch varieties, stating "I know of nothing more profitable that I could plant in the ground than a good variety of almond trees such as these are."[27]

Profitability, for Hatch, was not merely a point of personal pride, it was essential to the racial project of recruiting white settlers. Hatch promoted the Nonpareil along with other paper-shell varieties at state fairs across the country, winning blue ribbons and using his acclaim to broadcast an image of California's bountiful economic opportunity. Records of his phenomenal profits and images of his "mammoth" orchard, the largest in the state at the time, were featured in magazines and newspaper articles to entice settlers westward. One such article included an illustration of Hatch's vast plantings with the caption, "A Representative Orchardist—Rare Opportunities for Investment."[28] He personally toured Midwestern states, using his success in almond cultivation to woo prospective settlers. When Hatch lobbied the federal government for tariffs that would shield California's almond industry from competition, he explained, "If we could always command a market, a better class of immigrants could be attracted."[29] Profitable farming was a key strategy for ensuring white populations remained dominant.

The shift to paper-shell varieties is particularly remarkable given that many almond merchants in the eastern cities said consumers strongly preferred the familiarity of the hard-shell varieties imported from Spain. They requested that California orchards stick to hard shells. But California growers could not seem to produce them as profitably. Compared with peasant producers in Spain, California farms faced high costs given their reliance on wage laborers and cumbersome, monopoly-controlled railroads. Paper-shell varieties were seen as essential for competing with Spain's clear advantages: family labor and economical transport by boat. By the 1920s California growers had quickly saturated the limited local markets for almonds where population levels remained relatively low, and they were desperate to lure buyers away from the cheaper European imports. Paper-shell almonds gave them a crucial advantage, but ultimately the federal government would also kick in tariffs to protect almond growers' "landed investments."[30] In this way the state's trade policies worked in conjunction with the promotion of Hatch's new varieties to ensure the stability of Anglo farm ownership.

It is perhaps telling that while Hatch was touring the country to recruit white settlers, the labor on his own farm was almost exclusively performed by Chinese workers living in cramped quarters and frequently threatened with harassment.[31] The irony was not lost on white working-class organizers who demanded Hatch discharge the Chinese workers, asserting that their presence suppressed wages. Instead, Hatch doubled down. As a chronicler of local elites recounted, "he simply provided himself with a double-barreled shotgun, and putting it in the house where it would be handy, went about his business. He had been getting along with as few Chinamen as possible, on the principle that they were not a desirable element in our community; but threatened by the agitators, he told them boldly he would not suffer compulsion; he knew his rights, and would thereafter hire none but Chinese laborers."[32]

As a businessman, Hatch intended to maximize profits through the exploitation of poorly compensated Chinese labor, while simultaneously using his financial success as a recruitment tool for propertied white settlement. He even went so far as to use his continued employment of racialized workers as a symbolic gesture of defiance against labor organizers. Chinese workers were, as historian Alexander Saxton describes, the "indispensable enemy,"[33] eternally demonized and yet essential to wealth accumulation.[34]

Given his reliance on Chinese laborers, it is perhaps more appropriate not to solely credit Hatch with the "discovery" of the paper-shell Nonpareil. Chinese field workers walking the orchards day in and day out were no doubt keen observers highly sensitive to the trees' subtleties. The Nonpareil emerged from an assemblage. The interactions between an Anglo settler's racial, capitalist ideology; the almond seed's reproductive gambling strategy; the unique soils and climate of the Suisun Valley, newly bereft of its Indigenous inhabitants; and the attentiveness of Chinese orchard workers collectively brought it into being.

The Paper-Shell's Chemical Cost

The very quality that made paper-shell Nonpareil almonds so attractive to California growers would become their greatest vulnerability. Birds, insects, and fungi all found rich almond meats more readily accessible with thin, weakly sealed nuts.[35] Early growers of the paper-shell varieties that Hatch popularized noticed that birds would easily crack open such thin-shell nuts. A Santa Barbara grower reported, "The only varieties that have been tried in southern counties are Languedoc and Papershell. The latter has borne well but the birds get all."[36] A 1915 survey of almond growers in Butte County reported that woodpeckers were carrying away paper-shell almonds, with the caveat that "these birds are becoming scarce however as the old oaks are being removed."[37] Settler land clearance thus promised to compensate for the variety's vulnerability. Nearly eighty years later, a 1993 survey found that California almond growers were commonly shooting at crows or using propane flame canons to deter birds, yet most saw only a slight effect.[38]

A more significant consequence of the Nonpareil's paper-shell was the intrusion of smaller, more pernicious insect and fungal pests. In 1937, a pomologist working for the California Agricultural Extension Service reported, "The chief defect of the Nonpareil is that owing to its paper shell it is especially subject to the depredations of birds. For the same reason in many sections it has been infested recently by the peach worm. It is probable that spraying will have to be resorted to in order to keep down the pest."[39] The peach twig borer (*Anarsia lineatella*) wreaked havoc on orchards, and regular applications of lead arsenate became standard practice.[40] Soon thereafter, the rapid spread of the navel orangeworm (*Amyelois transitella*), moving north from the fruit districts of Arizona, reached California and established itself throughout the fruit- and nut-growing districts of California by 1961.[41] Early experiments revealed that "only soft-shelled varieties were infested."[42] The problem was not only the thinness of the shell but also its tendency to split open at much higher rates.[43] Mechanized harvest machinery made the situation worse by exposing the nuts for longer periods of time.[44] Knocking residual "mummy" nuts from the tree after harvest and destroying them is the surest known method for navel orangeworm control, but the high cost of labor and difficulty of ensuring neighboring orchards remained equally vigilant meant most growers also relied on pesticide sprays. Researchers found that the navel orangeworm and peach twig borer worked in tandem, the latter providing pinhole damage which the orangeworm could then exploit. Thus, to prevent costly navel orangeworm damage, farm advisers recommended thorough pesticide treatment of peach twig borer too.

The two pests combined provoked a series of pesticide treatment regimes whose legacy effects reverberate in ongoing environmental toxicities. Lead arsenate, a waste byproduct of copper smelting profitably redirected to agriculture,[45] accumulated toxic levels of lead and arsenic in soils, resulting in widespread contamination of orchard districts, which persists as a human health risk today.[46] When lead arsenate was eventually phased out and banned, it was largely replaced by DDT, a broad-spectrum insecticide eventually banned in the United States in 1972 after Rachel Carson's *Silent Spring* drew attention to its perilous impacts on wildlife.[47] Later research revealed DDT as a "persistent organic pollutant,"[48] more colloquially called a "forever chemical" for its resistance to degradation, and its enduring impacts on human health are still being uncovered.[49] With the banning of DDT, a class of chemicals known as organophosphates, originally developed as deadly nerve gases during the Second World War, then took the leading role in almond pest control. This potent set of tools is less persistent in the environment, but, unsurprisingly, it can permanently damage the nervous systems of those who experience acute or low-level chronic exposure.[50] Like DDT, its widespread use has made it one of the most common synthetic chemicals detectable in water, soil, air, plant, and animal (including human) tissues worldwide.[51]

For the moment many organophosphates are still legal in the United States and Europe, though a select few have been banned, and activists are mobilizing to gain regulatory recognition that their human toll is inexcusable.[52] Responding to mounting public concern, almond growers in California have gradually reduced the application of organophosphates since the mid-1990s. The trend has also been highly contingent on the availability of an alternative: synthetic pyrethroids.[53] Pyrethroids are much less harmful to mammals, but they are far from neutral. This class of chemicals is highly toxic to aquatic invertebrates and to fish, with implications rippling up the riparian food chain.[54] Like most insecticides, pyrethroids take away the "good" with the "bad," eliminating insects beneficial to agriculture and the broader ecologies of which they are a part. In humans, cases have revealed the potential for damage to neurological, respiratory, and reproductive systems, with particularly worrisome effects from prenatal exposure.[55] The transition from organophosphates to pyrethroids reduces the risk to human and broader ecological well-being significantly;[56] but it is a shift in degree, not in kind. The dynamic is perpetually unstable.[57] As the almond boom expands and intensifies acreage, the navel orangeworm and peach twig borer are finding abundant and contiguous habitat, evolving resistance to established treatments, and proliferating.[58]

Thin shells and their susceptibility to insects have cascading effects that intensify pesticide dependence. When insects penetrate the minimally protective layer of the soft-shell almond, fungal spores are free to enter. Two common molds (*Aspergillus flavus* and *Aspergillus parasiticus*) present in orchards produce aflatoxins, fungal carcinogens that almond processors take great pains to eliminate in order to avoid trade restrictions or a public health (and public relations) nightmare. While spores have many possible pathways for entering the almond, the most common by far is through insect damage. In fact, "several publications have suggested that almonds free of such damage are not contaminated by the toxin."[59] After harvest, piles of almonds awaiting processing are routinely fumigated to reduce the presence of aflatoxin and eliminate any remaining insect pests. For decades, methyl bromide was the fumigant of choice.[60] This potent, odorless gas is highly toxic to human bodies, putting farm and packing workers at risk from acute and chronic exposures if safety protocols falter. Following the global ban of methyl bromide for its role in ozone depletion, phosphine fumigation became standard practice. Phosphine is a flammable, highly toxic gas that is similarly dangerous for workers if any mistakes are made.[61] Its effectiveness diminishes as pest populations develop resistance, leaving the future uncertain.

The paper-shell Nonpareil almond championed by Hatch in the late nineteenth century drew almond growers into over a century of deepening chemical warfare with insects and fungi. The chemical weapons utilized to protect fragile Nonpareil almonds travel well beyond the farm gate. They follow workers to their homes, course through riverways and underground aquifers, and float through the air toward schools and neighbors. They are metabolized by countless other organisms, sometimes decades or possibly centuries after being deployed to save a fraction of a single year's harvest. Under uncertainty, public university extensionists have recommended pesticides for almond growers "as a form of risk management for both the grower's crop and for their own reputations."[62]

In fact, the state was involved at an early stage in encouraging the adoption of paper-shell varieties. Through farm outreach, university researchers have actively worked to persuade growers such that they might anchor rural communities and enhance the nation's role in global trade. Upon the retirement in 1952 of renowned almond agronomist Milo Wood, a report to the US Congress detailed his chief accomplishment as "the elimination of unprofitable and miscellaneous varieties. Mr. Wood's first assignment in the industry was a survey and study to indicate the most desirable varieties from the standpoint of the trade that were being produced in California and to conduct a campaign to convince growers of the advantages of them producing these most profitable varieties with the result that whereas when this survey was made some 35 percent of the almonds grown in California were of miscellaneous or poor varieties, today less than 3 percent are of miscellaneous varieties."[63] When individual California almond growers failed to optimize for profitability, the state stepped in to grease the wheels. In the early 2000s an advocacy coalition for reducing pesticide dependence chose not to pursue funding from the public university system because they saw the institution as advancing a purely economic agenda that sat in direct opposition to ratcheting down agrichemical usage.[64] The state has been central to the twin projects of homogenizing plant selection and chemicalizing agriculture throughout.

There was nothing predestined about the primacy of the Nonpareil. It was selected more for its suitability to political economic conditions than ecological conditions. The Nonpareil's highly profitable, chemically sustained fragility is in many ways a reflection of the fragility of the settler colonial project: thriving on expansion at an unstable frontier, propped up by racialized labor exploitation, and damaging to the very elements necessary for its success. Its precarity is now embedded in the landscapes and lives of California's Central Valley. Hatch's pursuit of almond profitability was inextricably tied with the instability of settler colonialism. Embedded in that paper shell are the aspirations for attracting and stabilizing white land ownership and a dependence on toxic compounds to sustain it.[65] The ongoing acceptability of chemical controls despite their toll on the bodies of farmworkers, along with their children and neighbors,[66] suggests that in agronomic knowledge of plant-pest relations, echoes of racialized hierarchy endure.

SPAIN'S DESMAYO LARGUETA:
PREACHING RESILIENCE TO AVOID REVOLUTION

The origin of Spain's most agenda-setting variety stands in stark contrast to California's Nonpareil. As Spain marked the end of its colonial era at the turn of the twentieth century, intellectuals turned toward the country's rural hinterlands to reimagine nation building rooted within their own territory rather than through overseas expansion. It was during this period that a priest in the remote Spanish village of Alquézar began breeding a weather-resilient almond variety. He was driven by a mixture of piety and politics, aiming to stabilize the livelihoods of those struggling at the societal margins and provide just enough propertied investment to suppress revolutionary sympathies. The physical roots of a long-lived tree like almonds, according to his vision, could help reinforce the established social order that he and his conservative contemporaries were anxious to protect. He chose almonds specifically because they demanded the least effort and resources to sustain. This botanical frugality was essential to his paternalistic, bootstrap message: When faced with poverty, trust in the advances of plant science to do more with less. Given this context, he approached his project privileging thrift and consistency over pure profits. While the Desmayo Largueta never gained in Spain the ubiquity of California's Nonpareil, it remains the most persistent named variety, and its objective set the course for nearly all Spanish almond breeding thereafter.

Rafael Ayerbe Castillo, known as "el cura de Alquézar" (the priest of Alquézar), gave his first mass in September of 1895 in a rural hamlet of the Huesca province of Aragón. The congregation assembled in the Colegiata de Santa María la Mayor, a medieval stone-walled church built atop a ninth-century Islamic fortress strategically perched on a cliffside overlooking the Vero River. Trained in botany as well as theology, Ayerbe complemented his religious duties with investigation into almond varieties. With substantial funding from the church or other personal sources, he hired workers to terrace and maintain an experimental orchard on the steep slope behind the Colegiata.[67] By stabilizing almond productivity, Ayerbe hoped to lift the incomes of the rural underclass and, through investments in tree care, affirm their adherence to structures of property and privilege.

Ayerbe's career was marked by a time of great uncertainty for Spain. Just three years into his tenure, Spain fought and lost the gruesome 1898 war against the United States in Cuba and the Philippines, surrendering nearly all of its overseas territories and ending its status as an imperial power. The year marked a crisis of self-image among Spain's intellectual classes, sparking a moment of collective soul-searching known as *regeneracionismo* (regenerationism).[68] While wide ranging in its political prescriptions, the movement's leading voices shared an emphasis on scientific progress and investments in rural spaces neglected by centuries of colonial expansionism. Foremost among their concerns was the destitution of rainfed farming areas, seen as a paragon of national deficiency and shame.[69] The goal of many reformists was to confront this entrenched poverty through hydraulic infrastructure, education, and science, while avoiding the instability of revolution. Fear of social upheaval intensified in the wake of Russia's 1917 Communist Revolution. Like many conservatives of the moment, Ayerbe's efforts to reform farmers through improved almond cultivation advanced a vision of entrepreneurial individualism that would remain loyal to the existing social order.

In the early twentieth century, Spain's capitalist agricultural economy was underwritten by a deeply entrenched and hierarchical rural social structure. Many rural residents were landless—forced to rent land, sharecrop, or work as wage laborers for larger landholders while navigating the influence of local bosses, known as *caciques*.[70] The land tenure systems varied widely between regions; Andalusia in the south featured large *latifundia* estates, whereas in the north, where Ayerbe was working, histories of parcellation had created dense patchworks of small *minifundia*. Both systems generated severe economic struggles for those lacking stable or sufficient land tenure, meaning that many rural households cobbled together an array of activities beyond working their own patch of land to make ends meet. Profits flowed to the processors, merchants, and landholders, and peasant producers were rarely able to make a living off their own land. When Ayerbe was tending to his almond experiments, great fears over potential land reform were brewing. The most well-remembered of the *regeneracionistas*, Joaquin Costa, came from a town just forty kilometers from Ayerbe's church and published an 1898 book titled *Colectivismo agrario en España* (Agrarian collectivism in Spain). In it he proposed socializing all arable land, to be managed by the state for the direct benefit of agricultural laborers, or at the very least a legal mechanism to prevent the collection of unearned rents. This was something the church, an ally to elites and a large landholder in its own right, could not abide.

Instead of refashioning the distribution of land, Ayerbe encouraged struggling *labradores* (peasant farmers) and *jornaleros* (day-wage laborers) to plant almond trees and to heed the agronomic guidance of scientists like himself in order to improve their lot. His writings expressed deep concern about the sway communist ideology might have for Spain's rural poor and hope that investments in the land would be an antidote to its allure.[71] In his treatise on the Desmayo almond, Ayerbe frames his ambition as a civic-religious-social education upholding the "principles of authority," even going so far as to recommend the use of force by the Guardia Civil (national police) as "the only remedy against those who in their inner sentiments do not recognize the legitimate legality of the right to property."[72] While Ayerbe concerned himself with the betterment of the rural poor, his dedication to almond research was part of a bootstrap mentality that upheld the fundamental structures perpetuating systemic poverty.

Ayerbe's concern for the plight of the poor blended piety with pragmatism; those with nothing to lose were the most likely to be radicalized. After receiving feedback that his original treatise was "too much reading for peasants and too expensive for day laborers," he published a brief, pragmatic summary for one sixth of the price of the original text. He begins the abridged booklet describing the dire state of rural affairs: "The lack of trees is the ruin of the farmer. The destruction of our olives and forests carries misery to our regions and families; emigration rising daily is an effect of this misery, and the misery comes from the lack of harvest, and what can one harvest when there are no fruits to gather?" He goes on to describe the predominance of wheat, with variable harvests under rainfed conditions, as costly in labor, fertilizer, and seeds. "The almond is the most resistant," he argues, "and for that they call it the king of rainfed farmland (*el rey de secano*): the tree that produces most and that requires the least of its owner."[73] In his efforts to breed the Desmayo almond, Ayerbe explicitly aimed to require minimal inputs, including both time and capital. The goal was not to maximize wealth but to give the restless rural poor a reason to stay put.

The Desmayo's distinguishing feature was improved protection from frosts. Almond blossoms emerge in winter (late January to mid-March depending on the local climate), exposing themselves to adverse conditions at the most tender moment in the tree's annual cycle. For many varieties, a drop in temperatures below 28°F (−2°C) for just thirty minutes can wipe out 50–90 percent of open flowers and embryonic nuts.[74] Farmers in the Mediterranean have long accounted for this sensitivity to frost by planting almonds above the valley floors where cool air settles, avoiding windy locations, and rarely relying on such a fickle crop for their primary source of income. The Desmayo (literally translated as "fainting") takes its name from the drooping position of the flower, which in hanging downward prevents moisture—and thus frost—from forming within its cuplike blossoms. Ayerbe marketed the Desmayo as "the least sick," "the least demanding of fertilizer and cultivation," and "the least likely to freeze and most resistant to cold."[75] Rather than touting its productivity or profitability, Ayerbe assured peasant farmers that the Desmayo reduced their risks by resisting disease, minimizing upfront costs, and limiting frost damage.

Ayerbe's prized variety proved its worth by growing in the most rugged conditions. His experimental orchard had very little topsoil and was planted on such a steep grade that without carefully maintained terraces it would be impassable. There could be no more of a stark contrast with the visual impression of Hatch's flat, manicured investment portfolio of plantations. Ayerbe's demonstration orchard was visited by an audience valuing the rugged character of a plant that can thrive despite the most challenging terrain. His almond trees were a means of grasping for economic stability, not a vehicle for amassing a personal fortune.

Ayerbe further recommended landscape diversity for economic resilience, suggesting that almond trees serve as a complement to olives and plums. He specified that the recommendations of his booklet, which minimize upfront investments, are fit for those without secure land tenure. His booklet recommends planting almonds on steep hillsides, on rocky or shallow soils, along edges of fields, or in spots where grains have been less productive. He even suggests planting them along public highways where "their riches could alleviate the poverty of road laborers."[76] The Desmayo was bred not to recruit settlers with idealized conditions and the leisure provided by hired hands but to persuade a downtrodden rural peasantry that they should put their faith in the authority of science, the church, and private property. Rather than portraying almond trees as engines of wealth, Ayerbe describes them, somewhat patronizingly, as a source of refuge. "Remember that your best friends, your richest protectors and your most capable and complete defenders are trees, and among them 'el rey de secano,' the Desmayo almond."[77] With a strongly resilient almond variety, one could plant their way out of poverty, or at least shake any notions of dismantling private property. Just three years after publishing his treatise on the Desmayo, the Spanish King Alfonso XIII awarded Ayerbe a medal of honor.[78]

A Legacy of Resilience

Resilience has become a popular framing for environmental protection, but it can also align with social conservativism.[79] Ayerbe's work exemplifies this tension. On the one hand, prioritizing self-sufficiency led to a path free of toxic materials. On the other hand, such thrift aimed to subdue more radical dreams of agrarian equality. The Desmayo almond serves as an artifact of this intersection, a reminder of the complex ways matter and meaning are bound together.

The thrifty and resilient Desmayo was widely adopted, by one estimate accounting for 95 percent of nursery sales in 1941,[80] and remains one of the leading named varieties in Spain.[81] The emphasis on resilience, to both frosts and pollinator interactions, continued as a central priority for a small but dedicated group of Spanish almond breeders in the late twentieth century (discussed in greater detail in chapter 3). In other tree crops, like citrus and olives, Spain has developed an intensified agro-industry similar to that of California. Yet almonds have largely remained a small-scale household enterprise, and breeders have, until the sudden boom in irrigated plantings, aimed to serve those working the land at the social and ecological margins.

Like all Spanish almond varieties, the Desmayo has a hard and well-sealed shell. Many almond growers in rainfed regions today credit this hard shell as a major factor in their ability to operate with minimal or no pesticide treatments. Often rainfed farmers I interviewed, especially those at higher elevations, would tell me they couldn't recall the last time they had a significant insect problem. A rainfed grower in the high plains of Andalusia said, "Here I've never used an insecticide treatment," including organic products.[82] Another farmer I spoke with who had recently converted her grain, almond, and sheep farm from conventional to organic said that very little had changed when it came to pest management for almond trees. A Valencian grower explained, "The hard shell. You [in California] have 60 percent output. Here, no. We're talking about a third lower. . . . It's a drawback because it has less profit, less value added, but it is a guarantee of the purity of the product inside. Why? Because the rate of contamination is zero." In fact, large swaths of rainfed orchards are considered organic by default, with growers using no synthetic pesticides or fertilizers but not bothering to obtain official certification.

As almond orchards, both irrigated and rainfed, have expanded in recent years, new pests of concern are emerging in Spain. But as one rural magazine article remarked in 2011, "if we look for technical information or science on the topic of pest control in almonds, we realize that there's very little information available."[83] Spain's new class of intensive almond growers are now challenged by the fact that there is scant pest control research to inform their practices.[84] As a crop of choice by those at the economic margins, almonds have never been given scientific priority in Spain, and as a tree selected to endure at the ecological margins without irrigation or fertilizer, pests simply have not been a major concern.

Given the hard shells protecting almonds from pest damage, the fungal spores of aflatoxin have not been a concern for the Spanish almond industry. An additional level of protection comes from the fact that nuts shaken from Spanish orchards typically do not touch the soil, a major source of contamination. When harvested by hand, growers use mallets and a blanket covering the ground. Since the 1990s most growers use an umbrella-like harvesting machine that wraps its mechanical arms around each tree trunk and vibrates to shake the nuts loose. These machines are deemed uneconomical for California's density and scale of orchards. Thus the combination of hard shells and small-scale operations has made aflatoxin a nonissue for Spanish almond growers.

When I interviewed Spanish almond breeders, meeting with leaders in the field across all major institutions whose work spanned the 1970s to present day, I asked if thin- or paper-shell traits had ever been considered for their breeding experiments in order to improve farmers' profits. The response was an unequivocal no. Hard shells fit the needs of most Spanish producers for several key reasons. First and foremost is protection from birds and pests. Soft-shell varieties are simply out of the question for those who could not afford time-consuming manual control measures or expensive chemical treatments. One researcher illustrated the extreme levels of bird damage by explaining, "If we want in our collection samples of Nonpareil, we have to go harvest them when they're green because if not, they will all have disappeared." Another breeder at a separate institution confirmed, "Our breeding programs don't want soft shells. Soft shells are a problem when the almond is damp or goes bad. We would have problems with aflatoxin. . . . There's the navel orangeworm." For rainfed growers such vulnerability is highly impracticable, not only in terms of upfront costs but also temporal flexibility within a diversified system. Unlike soft-shell varieties, which must be processed immediately to avoid spoilage, hard-shell almonds can be collected and set aside while more perishable crops are harvested and sold. A grower and local nut industry coordinator from Aragón explained, "The Spanish almond has a much harder shell than the American. It can handle two or three years without being shelled." Thus these varieties are better suited to a diversified farming system. The protection of hard shells means nuts can be banked for times when other tasks have dwindled, when market conditions align or when cash is tight. Hard shells also make it easier to remove the hull (the soft, fleshy outer layer of the fruit) with simple manual household machinery that many rural Spanish households possess. By remaining less reliant on time-sensitive processing by a third party, growers retain more autonomy. In keeping with Ayerbe's guiding priorities for the Desmayo, hard-shell almond varieties are valued for reducing risk, limiting maintenance, and accommodating the diverse economic activities of marginalized rural households.

BREEDING MATTER AND MEANING

The foundational almond varieties in California and Spain took radically divergent trajectories with undeniably political underpinnings. Patterns of plant observation, characterization, and selection have been shaped by ideological priorities. These priorities have produced ripple effects generations later, material consequences that are inextricable from but also far surpass the breeder's intent. While A. T. Hatch may never have imagined the pesticide dependencies of today's intensive almond orchards, we cannot fully understand the centrality of the paper-shell almond and its associated chemical cocktails without accounting for its role as an instrument of racial capitalism and Anglo settler dominance. While Rafael Ayerbe may never have considered that almond varieties selected for resilience could minimize toxic exposures, his desire to root peasant farmers in place is encapsulated in their forms. The materiality of an almond tree reflects layered histories, a fusion of intentionality and contingency.

While the outcome of the Spanish case is undoubtedly more desirable, both ideological projects are deeply problematic. This provides a healthy reminder to temper romantic assumptions about what underlies a system with admirable characteristics. It also cautions against a simplistic inclination to seek out narratives of innocence.[85] Resilience may lend ecological stability, but it does not inherently question structural oppression. If we cannot disentangle the Nonpareil's pesticide treadmill from California's settler colonialism, then we cannot disentangle the Desmayo's resilience from Spain's conservative rural politics. Where does this leave us?

First and foremost, the juxtaposition raises important questions about how things might be otherwise. Would insecticide sprays and toxic vapors be so crucial to the California almond industry if growers produced nuts with a stronger protective shell? Almost certainly not. Breeders and agronomists are well aware of this, explaining in our conversations that as pesticide regulations become more restrictive, a shift to harder shells may be warranted. As early as 1976, a renowned almond breeder, Charles Grasselly, lamented the path dependency set by the naive choice of defenseless paper-shell almonds. "It would be preferable to go back to the more resistant shell types," he concluded.[86] The few organic growers in California (accounting for only 0.03 percent of all California almond sales in 2017)[87] already rely on hard shells to provide protection without synthetic pesticides.[88]

The Nonpareil's weaknesses were evident very early on, and pesticides were quickly proposed as a solution. Why were such costly remedies readily adopted? One key reason is that almond production in California was dominated by farmers with significant capital. In many cases, surplus capital from mining and related industries was the raison d'être of the farm itself. With hard-shell varieties as the status quo, farmers of meager means would have balked at the notion of choosing a vulnerable variety in order to make a high-stakes bet on future returns. It is a bet that only makes sense for someone like Hatch who sees in the plantation-styled farm a strategic wealth-multiplication machine. Hard-shell almonds would likely have been a fine choice for more modest economic ambitions. Soft-shell almonds like the Nonpareil were deemed necessary only to sustain eye-popping profits that could attract privileged bodies for expanding racially stratified settlements.

It is illuminating and encouraging to see that almond agriculture need not have followed California's path of toxic dependencies. To see that the hard shell offers a simple path away from chemical toxicities and yet the pesticide-dependent paper shell continues to dominate is astonishing and downright infuriating. What was even more frustrating for me as I spoke with farmers and agronomists in California was how easily they disregarded the idea of promoting hard shells due to more limited profitability. The problem was not available "technology" but a political economic context that precluded an obvious and pragmatic possibility.

A second major insight drawn from these parallel cases is the significance of the material conditions of production, namely land and labor, to agricultural knowledge practices. Hatch sought to displace Chinese laborers with white settlers. Ayerbe sought to teach peasants that they could eke out a better living without resorting to land reform. Who was intended to farm, under what type of labor and land configuration, profoundly shaped breeding priorities. The lesson to be drawn from this is not that environmental protection requires Ayerbe's conservative politics. Far from it. Rather, seeing the significance of land and labor means that we cannot ignore their role in ongoing knowledge production processes, in plant breeding and beyond. What does it look like when we turn the same lens on contemporary agricultural sciences? What assumptions about land and labor are just beneath the surface when hazardous pesticides are deemed essential, or when a switch to hard-shell almonds is brushed aside?

Today's agricultural knowledge practitioners might balk at the notion that they carry an ideological agenda. It's easy to look back at Hatch and Ayerbe as relics of another time when promoting white supremacy or protecting landed wealth was socially acceptable or even expected of prominent agricultural thinkers. But such explicitly oppressive intent is not required for matters of land and labor to lie just below the surface in agronomic inquiries. Choices made about what plants need, including what they need to physically be, have clear political economic foundations and consequences. As with the almond's paper shell or drooping flower, these matters are woven into matter itself.

2

———

Flow

Knowing Plant-Water Relations

At a distance, almond trees appear as a sprinkling of green dots against a background of pale yellowish-gray soil, scattered across hillsides throughout much of Mediterranean Spain. Sometimes terraced along contour lines, sometimes stretched across arid plains in neat rows, sometimes snaking their way along roadsides or property lines, in this place almonds are often called a "crop of convenience," easily integrated into rural interstices. The width of their spacing is evidence of their reliance on rainfall. The drier the climate, the more room each tree is afforded to extend its roots throughout the soil. Among rural Spanish households, almonds have long been famed for their ability to make the best of adverse conditions, growing and producing in rocky, low-nutrient soils with long, dry, and searingly hot summers. Given that most crops need some added water to withstand these conditions, irrigating an almond tree beyond its initial planting had, until very recently, seemed unthinkable. Best to save scarce water for crops where it is an absolute necessity. Almonds are expert at doing without.

Spain produced the largest share of the world's almonds this way, sparse rainfed trees across hillsides and interwoven with other rural activities, until its production was surpassed by California in the 1970s. California almond trees are intensively irrigated and currently yield nearly 80 percent of global production.[1] Throughout the flat, wide expanses of the state's agricultural heartland, the Central Valley, orchards form a dense carpet of deep green. The continuous lush tree canopy is broken at orderly intervals by stripes of pavement or pipes. Groundwater pumps rise from intermittent blocks of bare earth where industrial tubes lift and redirect water from aquifers deep below. Some farms are bisected by a stripe of vibrant blue, channelized canals carrying snowmelt from distant mountains.

48

FIGURE 3. Rainfed almond trees in the foreground and in the distance along property boundaries in Aragón, Spain. Photo by author.

FIGURE 4. An intensively irrigated almond orchard in Kern County, California. Photo by author.

A fine capillary of black plastic tubing siphons water from below and beyond, toward the thick mat of roots sustaining the verdant forestlike fields. Almond farmers utilizing intensive irrigation in California, and increasingly in Spain, are applying more water per unit of land than ever before.

So how much water does an almond tree actually require? From an ecological perspective, the plant needs just enough water to sustain its reproductive success. Generating good quality seeds that can find suitable soil and perpetuate the species would do. In an agricultural context, however, a plant "requires" what farmers expect from it in order to sustain their livelihoods, or increasingly, to maximize revenues for whomever owns the land. As the self-proclaimed "almond doctor," David Doll explains in an article for the industry magazine *Growing Produce* that "although almonds can survive on as little as 7.6 inches of water annually, mature, high-yield orchards in Kern County [California] require as much as 56 inches of applied water."[2] Fifty-six inches is exactly what it sounds like: an acre of farmland, or tens of thousands of acres, with 4.67 feet (1,422 millimeters) of standing water. That is the volume of water applied per year to almonds in Kern County, a region that receives on average about six inches (152 millimeters) of rainfall annually.[3] Almonds evolved in Mediterranean-like climates characterized by hot and dry summers, and fifty-six inches of rain does not regularly fall in any climate on earth where almonds grow. Put another way, according to agronomists, almonds require more water than falls as rain in any place where they are physiologically capable of existing. How did a crop's water "requirement" become so divorced from its ecological context?

An almond grower living in California or elsewhere one hundred years ago would have assumed such an immense quantity of water would surely kill the tree. And they would almost certainly be right. The orchard had to be refashioned several times over in order to accommodate a deluge that would net dollars instead of damages. Almond trees are highly sensitive to having "wet feet" for too long and quickly suffer from water-related pathogens or root asphyxia.[4] Grafting almond trees onto peach rootstocks allows the trees to better tolerate moisture. Such an immense quantity of water can feasibly be applied only by a year-round trickle or mist by high-efficiency irrigation technologies, such as drip lines and microsprinklers. Only an orchard with trees packed in tightly to capture nearly every square centimeter of the sun's rays could make use of so much water. And for orchards battling the accumulation of salts, due to decades of heavy fertilization and "efficient" irrigation, fifty-six inches of water will likely be surpassed in order to push the salty layer deeper underground. The sheer volume of water is, as many Spanish farmers I spoke with stated simply, "*una barbaridad*," an outrageous quantity. There is nothing inherent to the almond tree about it. It is a miracle of modernity that the almond tree can withstand it.

Even California almond agronomists readily acknowledged in interviews that irrigation levels this high had taken time for growers and researchers to accept.

Farmers had to be painstakingly convinced by careful research codified into widely circulated irrigation recommendation formulas. Gradually almond "crop water requirements" across the California industry crept higher and higher. When the almond boom hit in the early 2000s, these formulas jumped scales, as agronomic science was exported to new extractive zones in Spain, as well as Australia. This was not a clear or inevitable path for the almond. It required the convergence of specific orchard forms, technologies, geographies, ownership structures, and knowledge practices, all advanced by an agrarian capitalism deeply intertwined with the state via public universities, water infrastructures, and networked evapotranspiration measurement stations. Almond trees, long cherished for their drought tolerance, are not "thirsty." The thirst for steady profits has refashioned ways of knowing the orchard's flows.

When it comes to irrigation, rather than major innovations in plant genetics,[5] almonds have experienced innovations in agronomic knowledge. By this I mean not that scientists and farmers know more about almond irrigation but that they come to know it differently, through a distinct set of mechanisms and processes. Chronic overproduction, combined with the rising influence of corporate farms working in tandem with public land grant research institutions, have fundamentally reshaped almond irrigation knowledge. They have redefined how much water almond trees "require," and in doing so they have exacerbated the almond boom and its attendant environmental consequences. Almonds provide an emblematic case of how knowledge politics shape agriculture because, as a long-lived tree, systemic data gathering and analysis take decades and requires extensive resources. Capital and state infrastructures work to not simply reveal the underlying nature of almonds but rather actively produce agronomic facts codifying extractive logics.

In this chapter, I aim to denaturalize the "crop water requirement," a seemingly neutral mathematical calculation used not only in the almond industry but all modernist agricultural projects. While the term and its significance may seem merely descriptive, the ascendancy of the crop water requirement to a position of unquestioned fact is very much political. The conditions required to calculate it, and the assumptions made to support the calculation, accept the unrestricted use of agrichemicals and a network of carefully calibrated, state-funded, real-time monitoring stations. Even the mathematical constant used in the calculation itself has crept upward over time to reflect intensification of the production system. Each new form of intensification has become codified in a rising "crop water requirement," the basis of deciding not only how much water flows into the orchard but also how much money is expected to flow out. Scientific assumptions quickly become economic expectations. Irrigation scientists describe the "crop water requirement" as simply a physiological feature of a plant provided "optimal" conditions, calculated without regard for local hydrological cycles and water availability. This extreme isolation of a plant from its ecological relations is an elaborate fantasy, sustained by the unsettling premise that the source of water and its associated relations do not matter. For the communities of humans and other beings who lose their rivers' vitality to dams and diversions and struggle to sustain themselves as aquifers run dry, such conditions are a far cry from "optimal."

In addition to illustrating the political contours of the crop water requirement, I also emphasize here that two prevailing obsessions within irrigation science, efficiency and precision, have done little to reduce water use on a given plot of land. In practice, they've often had the exact opposite effect. Efficient irrigation technologies are often assumed to result in less water use, but for almonds they have allowed for the expansion of irrigation into arid lands that never would have been viable otherwise. Plant stress measurements are assumed to result in more restrained, and precise, water use, but they can also uncover new opportunities to increase yields. It turns out that the more you look for "stress," the more likely you are to find it and thus unlock productivity potential.

I juxtapose this fixation with plant stresses against the careful ways in which Spain's rainfed growers fixate on working with the plant strengths. Theirs is not a magical recipe for achieving intensive yields without irrigation. Rainfed growers have far lower yields, but also very low costs and a more modest, diversified livelihood. Without dazzling productivity figures or a desire to buy much in the way of agricultural inputs, their forms of agricultural knowledge are trivialized as inefficient or irrational. I use the comparison to serve as a reminder of the kinds of plant-water knowledge that take root largely on the peripheries of capitalist and state investments, out of the necessities of rural precarity rather than the drive for maximal return on investment. This diffractive comparison holds up a prism to hegemonic, universalizing irrigation science and exposes its situated politics.

The rising water consumption of almond trees is not solely a concern for those with a stake in the transformation of water worlds in almond-growing regions. It signals a dynamic likely taking place among crops experiencing similar surges in financial investment, corporate management, and state support, such as asparagus and pistachios, also expanding in arid lands.[6] Irrigation and intensification continue to expand worldwide.[7] Furthermore, once a crop is expected to require and utilize higher levels of irrigation, demand for fertilizer along with more complex pest management strategies rise to accommodate more dense growth patterns. New levels of astonishingly high yields then become benchmarks in valuing land.[8] Many of the most prominent almond farm managers I interviewed were also real estate consultants, assisting buyers (increasingly institutional investors like pension funds) by evaluating the expected productivity of purchases under consideration. Crop water requirements and the yield expectations they entail effectively bake extreme levels of extractive water use, and the associated agrochemical agents, into the value of the land itself.

FEARS OF OVERPRODUCTION INSPIRE IRRIGATION

Almonds, extraordinarily well adapted to dry environments, are perhaps one of the most unlikely crops to become intensively irrigated. In Spain, since at least the end of the Islamic period in Iberia (1492), almonds have served a complimentary role to the regional staples: wheat, grape vines, and olives. In this system, water for irrigation is supplied, when available, to small gardens (*huertas*) where diverse vegetables are grown or is sent through gravity-fed channels (*acequias*) to ensure the productivity of staple grains. Almonds planted along the borders of these areas or along channels as a means of stabilizing the soil absorb some water by virtue of their location, but regularly irrigating an almond tree beyond its initial planting is extraordinarily rare. Almonds have also been a popular choice for hillsides less suitable for grains, where they are protected from frosts which form where moisture settles along valley floors. These arrangements continue in rainfed regions to this day.

When Spain began establishing missions in California in 1769, some peripheral attempts at almond production indicate no irrigation used. In the wake of the California Gold Rush of 1848, when a surplus of capital was invested in agricultural land speculation, a new class of business-minded settler capitalists took up fruit production.[9] Despite a proliferation of irrigated districts for citrus or stone fruits, almonds retained their rainfed character. In Sutter County, very few growers irrigated almond orchards unless drought persisted, stating "many of the most prominent almond men have found irrigation is not advisable and that the trees are apt to be injured" by root damage due to lack of drainage.[10] In nearby Acampo, despite relatively low precipitation (14.5 inches, or 368 millimeters), the water table occasionally rose too high, causing the almond roots to rot.[11] Even the newly established university research plots testing varieties and pruning techniques received no irrigation whatsoever.[12] Irrigation would provide a competitive advantage during dry seasons, one researcher notes, though most orchards irrigated once or at most twice per year.[13] A 1925 report concludes that "where winter rainfall averages 16 inches, almonds can generally be grown . . . without irrigation" and cautions against the almond's sensitivity to high water tables in irrigated districts.[14] Among California's growing class of settler orchardists, almonds became a popular investment, but irrigation was understood to be largely unnecessary, used sparingly if at all, and even deemed hazardous.

How did California almond growers go from seeing irrigation as an opportunistic boost to an absolute necessity? The initial motivation can be found in the context of a settler colonial state, where a rush of newcomers hoping to find their fortune confronted a challenge of their own making: overproduction. As historian Steven Stoll writes, California's early orchardists were not long-time farmers looking for a humble, rural existence but rather a class of cosmopolitan investors determined to make farming a profitable industry to build their wealth.[15] But so many similar growers jumped into the business all at once, markets for their products became quickly saturated, and prices risked dropping precipitously. Almond growers banded together to form a cooperative in 1910, boosting prices by 50 percent, but then became victims of their own success. Strong prices inspired another planting spree, and they were soon teetering on collapse again. Total production in 1919 was double that of the year prior.[16] The persistent fear of declining prices drove immense investments in marketing (see introduction).[17] It also drove individual growers to increase yields by any means available. Convincing Americans to eat more almonds, which at the time were considered strictly as a winter holiday treat, seemed much more uncertain than fashioning new ways to boost production. If they couldn't control the price of almonds, at least they could compensate with higher output. Writing in 1918, a university bulletin to producers explained, "Within the next few years California growers will, in all probability, be forced to accept lower prices for their almonds. . . . It is essential therefore that a careful study be made of all the factors concerned in the growth, production and final disposition of the almond crop."[18] Unlike annual crops like grains or vegetables, long-lived orchards were a multidecadal investment.[19] A grower could not easily switch to a different crop. Best to ensure that in the event of low prices, yields would be high enough to balance the books. While this rationale made sense at an individual level, at an industry level both strategies to confront overproduction—marketing and yield increases—eventually made the problem worse.

Planting booms and their associated overproduction fears came in waves. After years of spectacular advertising campaigns and political lobbying for a protective tariff to stave off a price crash, almond prices took a dramatic dive during the Depression years of the 1930s, and plantings slowed.[20] They shot back up, however, when imports evaporated during Spain's Civil War (1936–39), prompting a new surge in plantings. When those trees came into full productive maturity about seven years later, total production jumped up by as much as 48 percent in a single year with prices sinking by 34 percent. It is during this time in the mid-1940s when yields show a step change, from consistently landing in the range of 180–240 pounds per acre to 430–550. The only clear explanation is the rapid expansion of irrigation. The boost in almond yields conspicuously coincided with the opening of the federally backed Shasta Dam in 1945, which directed flows to farms in the Sacramento and later San Joaquin Valleys.[21]

Irrigation does not require massive public infrastructure, as it has been widely available since the late nineteenth century through private "ditch" companies diverting tributaries and later via electric groundwater pumps.[22] But a dam makes it much cheaper and more accessible. To make the leap to irrigation, however, almond growers had to fix the vexing "wet feet" problem. Researchers and orchardists began experimenting with grafting almonds onto the rootstock of a peach tree, a close relative botanically speaking but adapted to far wetter climes. According to a leading almond breeder at the University of California, if almonds are California's "Cinderella crop," then peach rootstock is the glass slipper. It has enabled the proliferation of almonds into valleys where production is supercharged by fertile alluvial soils and abundant irrigation. Growers began treating the almond like a peach.

Without the switch to peach rootstocks, almonds are one of the last crops one might consider as requiring irrigation. Among fruit trees, they have an unparalleled capacity to alter their summer growth pattern to survive extended dry periods.[23] This summer semidormancy allows the tree to slow its vegetative growth and prioritize fruit production as water becomes scarce. As the breeder explained in our interview, "You don't see summer dormancy in almond if you put it on a peach and you pump it with water and nitrogen. You're just overriding that and you're just pushing it through." The same capacity that allows almonds to stretch small amounts of moisture over extended periods also means that the tree is highly responsive to water when it becomes available. Rather than taking advantage of the almond's exceptional capacity for drought resilience and productivity on marginal land, California growers have harnessed the capacity for peach-grafted almonds to flourish under intensive fertilization, irrigation, and suppression of the pests attracted to such abundance. This overdrive mode came to appear necessary in large part due to California's distinctively settler colonial circumstances. Almond growers were by and large not seeking a humble homestead but a place to sink capital in real estate with the expectation of a handsome profit. As investor orchardists poured into the state in the late nineteenth and early twentieth centuries, they had money to spend but not enough people to buy their products. An initial surge of surplus capital created a constant state of anxiety over the lack of markets. The yield boost from irrigation offered individual growers a sense of security in their investments, despite contributing collectively to the mounting problem of overproduction.

MICROIRRIGATION ALLOWS CROPS
TO BECOME THIRSTIER

Until the 1980s, nearly all irrigated almond orchards were "flood-irrigated." At discrete intervals, an ample volume of water would be diverted from a nearby stream, canal, or groundwater pump such that it could flow into the orchard and saturate the soil. Depending on local conditions, farmers often dug furrows in between tree rows, or borders along section edges, to allow percolation in the root zone without risking moisture-related damage to the trunk of the tree. For those with access to surface water, this simple system was incredibly cost effective. Beginning in the late 1970s, however, almond growers rapidly adopted new drip irrigation systems. While drip lines are commonly associated with water conservation today, the primary motivation for growers was profit.[24]

Drip irrigation, a network of tubing allowing slow and localized water application, was most prominently promoted in the United States as a yield-boosting technology.[25] Trees grow more vigorously with a steady trickle of water than with alternating wet and dry periods. Often the advantage is not exclusive to irrigation but is compounded by the advantages of a suite of intensive agricultural practices facilitated by "drip." The infrastructure of microscale water conveyance can also deliver more consistent and localized applications of fertilizer (fertigation) and pesticides (chemigation), with both yield-boosting and cost-saving impacts.[26] The drip revolution thus allowed for a substantial decrease in labor associated with agrichemical applications and an increase in plant uptake of input through slow and steady feeding. Microsprinklers, carrying the same set of advantages, later provided an additional option.

As a technology of efficiency, microirrigation enables the crop to take up more of the applied water, fertilizer, and agrichemicals than it would otherwise. More of these inputs contribute to the plant's growth and less dissipate in the surrounding environment. These gains, however, do not always translate into reduced overall applications. In fact, a phenomenon known as the Jevon's Paradox often arises, in which the cost savings of increased efficiency allow growers to use the resource in greater quantities or in regions where it was previously uneconomical.[27] Both scenarios have proven true in the case of almond irrigation. Growers began applying more water per acre than ever before, and orchards sprouted up in locations that would have been inviable without precision irrigation. In 1975, almond growers in the San Joaquin Valley (the drier southern stretch of California's growing region) applied just over thirty-seven inches of water.[28] After the adoption of drip and precision scheduling based on plant water uptake, this steadily swelled to fifty-six inches or more, rising by more than a third.

The diffusion of microirrigation technology and its effects was neither immediate nor inevitable. Converting growers took a great deal of effort on the part of public university extension workers and irrigation equipment manufacturers. Extension agents with the University of California engaged in research, managed demonstration plots, and enthusiastically promoted microirrigation by supplying documentation of its yield-enhancing and cost-saving benefits, tailored to specific crops and conditions.[29] Manufacturers readily donated equipment.[30] Given the high upfront cost of a drip irrigation system, growers with access to relatively cheap water could not justify the expense based on water savings alone. In fact, water savings have featured so minimally in growers' calculations that California orchardists who have converted to drip are known to switch back to flood irrigation if it maintains equivalent yields with reduced maintenance costs.[31] Almond growers were among the earliest to widely adopt drip because orchards had expanded gradually farther south into the hotter, drier San Joaquin Valley where evapotranspiration rates are higher and along the valley's western side where water is more expensive.[32] High prices for almonds also justified the investment in terms of increased yields. As of 2020 the California Almond Industry boasts 85 percent adoption of microirrigation.[33]

THE CROP WATER REQUIREMENT TAKES ROOT

Microirrigation's profusion not only transformed the orchard landscape but also the way growers thought about water. In addition to articulating irrigation as a matter of volume (most commonly "acre-feet"), they began speaking the language of "crop ET" (ET_c) or "crop evapotranspiration" in order to meet the "crop water requirement." Plants draw water into their roots to replace the water that is being transpired out of the openings (stomata) of their leaves as they photosynthesize. The plant consumes only a minuscule fraction of the water taken in to produce new tissue, as over 99.9 percent passes up and out into the atmosphere. The amount of water one expects to travel up into the atmosphere on a given day, or "crop ET," includes plant transpiration as well as evaporation from both soil and plant surfaces. Evapotranspiration describes how quickly water moves through the soil-plant-atmosphere continuum.

The calculation for estimating the crop water requirement (ET_c) on a given day is fairly simple. You multiply the known ET of a reference crop (ET_o), most commonly cut grass or alfalfa, by a crop coefficient (K_c), a constant that conveys the unique characteristics of a given plant at a given life stage:

$$ET_c = K_c \times ET_o$$

In 1975 the United Nations Food and Agriculture Organization (FAO), in partnership with a University of California irrigation researcher, began widely disseminating crop water requirement guidelines based on this equation.[34] These were not originally intended as a recipe for individual growers to follow, however. Rather, the FAO sought to assist the planning and proliferation of new irrigation projects around the globe. The equation was designed to calculate maximum possible demand for a given region such that engineers could design a suitable system or determine how much land could be served by a given project. The crop water requirement calculation aimed to inform dam builders, not farmers.

Given the goal of ensuring adequate supply during peak demand, it is understandable that crop evapotranspiration represents almost implausibly luxuriant conditions for a given crop. The FAO report defines crop water requirements as "the depth of water needed to meet the water loss through evapotranspiration of a disease-free crop, growing in large fields under non-restricting soil conditions including soil water and fertility and achieving full production potential under the given growing environment."[35] Thus the simplicity of the equation belies a complex set of underlying assumptions. The standard is the industrial ideal: a large, uniform monocrop growing in top-quality soil, provided ample fertilizer, and using any means necessary to achieve complete pest control. The crop is imagined, quite literally, as unlimited, free of all ecological relations that may inhibit maximal growth. While the authors of the report encourage "caution and a critical attitude" in order to adapt the methodology to suit local conditions,[36] its function is that of an upper limit. It is exceedingly generous in order to avoid peak demands that surpass the capacity of regional irrigation infrastructure.

It was not long, however, before microirrigation allowed agronomists to take the crop water requirements as guides for optimizing production. This type of "model transfer" that applies a formula to suit a new purpose is quite common in the sciences.[37] For agronomists, the goal of irrigating at the level of crop ET is to eliminate the possibility that water might become a limiting factor to crop productivity. Any less and the crop is considered in "deficit" and thus risks experiencing water "stress." A plant is considered to be in deficit when it is releasing water more quickly than it is absorbing it, a common daily occurrence. Stress means that a water deficit is causing the plant to change how it operates in order to conserve water for the most essential tasks. Exceptionally few plants receive water in such consistent abundance as full crop ET. Flood-irrigated trees experience significant periods of dry soil between waterings, somewhat akin to rain events. This was, after all, the major selling point of microirrigation. Constant water in small doses means the plant can take up not only a higher portion of applied water but, in pursuit of higher yields, more water overall.

Crop ET had been a familiar concept in agronomic research for decades, but it was not until the combination of microirrigation and meteorological measurement, both receiving generous state support, that it became common parlance among almond farmers.[38] Growers began relying on university-generated recommendations for daily irrigation decision-making to maximize yield. Nearly every farm manager or owner I interviewed explained that they simply followed the recipe from the University of California in terms of irrigation and fertilizer applications. Considering the relatively low cost of water for most growers and the rising value of almonds, slight changes in calculations for evapotranspiration could effectively move immense volumes of water into orchards throughout the state.

Precision capabilities meant water application rates could be fine-tuned according to local conditions on a daily basis. This is a task very few farmers could undertake on their own. An immense, public knowledge infrastructure was developed to support it. Calculating crop ET (ET_c) requires two data points, the ET of a reference crop (ET_o) and a crop coefficient (K_c). The reference ET varies depending on weather conditions including solar radiation, air temperature, wind speed, and relative humidity, as well as soil characteristics. For reference ET to be most accurate, these conditions need to be measured for the reference crop at a local weather station near the farms that will make use of its data output.

Beginning in 1982, the University of California sustained daily weather monitoring across the state through the California Irrigation Management Information System (CIMIS). At each of the now 145 sites, weather monitors are located above a meticulously managed landscape to ensure that the measurements are as standardized as possible. "The ideal site for a CIMIS weather station is a 20-acre [8 hectares] or larger cool-season perennial grass pasture that is well maintained. The station itself will be located in the center of the pasture, inside a 10-yard by 10-yard fenced enclosure. Inside the enclosure, the grass would also be well maintained (properly irrigated and fertilized) and mowed frequently to maintain a height between three to six inches."[39] In the hottest, driest desert or the meadows beside towering peaks, public employees ensure a landscape akin to a lush suburban lawn thrives uninhibited so that farmers statewide can have a benchmark for irrigation tailored to their local conditions.[40]

In theory, obtaining a reference ET by positioning a weather station above twenty acres of flawless grass is the hard part. The only other factor in the equation is the crop coefficient, K_c. The symbol for the crop coefficient uses the letter K because mathematically it functions as a constant. Hypothetically, anyone anywhere in the world could take the same K_c, multiply it by the reference ET that reflects local conditions, and calculate their crop water requirement. As it turns out, this constant has actually been changing over the years.[41] K_c has been gradually creeping upward, both reflecting and proliferating intensive irrigation practices. A higher K_c means higher crop water requirements, recommending greater volumes of water applied to a given stretch of land.

THE CONSTANT CREEPS UPWARD

When is a constant not constant? How and why did K_c creep upward over time? The original FAO document for calculating crop water requirements included tables detailing crop coefficients (K_c) for major crops grown around the world. These reference tables, compiled from the most prominent scientific studies, are imagined to be applicable in any location. But there are a number of assumptions baked into them. The constant is derived from careful measurements of a given crop over the course of its growing cycle. The management of the crop directly influences these figures. For example, a heavily pruned tree with less vegetative growth would have less water vapor transpiring from its leaves. A chart of K_c values thus represents not just the physiology of a given plant but also the cultural practices in place. Agronomists know that K_c values vary with agronomic practices.[42] The significance of a rising K_c for almonds is that it universalizes and codifies increasingly intensive production.

As California almond growers in the late twentieth and early twenty-first centuries intensified production, heeding the advice of university extensionists, their practices became reflected in a series of new studies revealing higher K_c values. Irrigation recommendations going forward presumed elevated production levels dependent on not only more water but also the heavy fertilizer loads and pest control measures accompanying high-density orchards. When high prices set off a boom in almond plantings, those new to the industry in California, as well as Australia, and irrigated regions of Spain were eager to follow the latest recommendations. These updated K_c values provided a recipe for almonds that "required" more water per unit land than ever before.

Almond-growing practices in California have shifted substantially since the first K_c values were widely circulated in the 1970s, when precision irrigation was first popularized. As a group of agronomists noted in 2012, "In general, these midseason K_c values were more than 25% higher than the 0.90 value that was used for decades, confirming that more realistic K_c values were needed."[43] The trees were being grown more densely and intensively than ever before. These were not necessarily shifts initiated by growers. Outreach through extension agents, at conferences, and through training of crop consultants and other industry professionals took considerable effort. This was the case not only for disseminating new irrigation standards, like K_c, but also for all the agronomic practices that resulted in the rise of K_c. Many dramatic changes, such as the nearly complete abandonment of pruning, shocked growers at first, before they were ployed with data to convince them otherwise. University researchers, often with funding or donations from industry, persistently discovered new levels of intensive production to maximize profits per acre and then worked diligently to disseminate their findings. In conversations, many expressed that academic work can be grueling and thankless, but the knowledge that they were directly helping farmers made it more rewarding. Their careers depended on new discoveries for publications and funding, but unlocking new ways to optimize growers' operations was also a means of finding personal fulfillment. In roughly two decades, research prompted four major shifts, all of which would effectively boost water use: more dense plantings, limited pruning, heavier fertilizer, and postharvest irrigation.

Tasked with boosting profitability for growers, researchers found that almonds can grow more densely than had previously been practiced. A group of extension researchers found that for every additional 1 percent of light intercepted by an increasingly dense canopy of leaves, an acre would produce an additional fifty to fifty-eight pounds of almonds.[44] According to one extensionist who disseminated the conclusions of this research, any sunlight that reaches through the canopy to the orchard floor is effectively a "missed opportunity." This mantra of fifty extra pounds for every 1 percent of light circulated widely among growers. (The caveat that trees at a density above 80 percent light interception might increase the incidence of food-borne illness,[45] less so.) More tightly packed orchards contribute to higher K_c values (and thus higher irrigation rates), as K_c is well-known to correlate with light interception.[46] Importantly, this yield boost for maximizing light interception is only the case if water is unlimited. Rainfed farmers have long understood that more widely spaced trees endure arid conditions by accessing a larger volume of soil moisture.

Growers could be convinced to plant trees a bit closer, but many were scandalized when researchers initially suggested they cut back on pruning or toss the shears aside all together. Almond growers in California had naively modeled their pruning practices on peaches, agronomists explained, where widely spaced branches are required to allow ample light and air to reach large tender fruits. Trials of almond orchards with limited to no pruning showed that the sizable effort of removing woody branches each year was not necessary. Orchards were not necessarily higher yielding without pruning, but there was little difference in yield with significantly reduced labor costs.[47] After initial efforts to structure the trees in the first few years, growers could allow the tree to determine its own growth pattern and make cuts only to remove disease, prevent wind damage, or accommodate equipment.

Concurrent with the shift toward higher density, minimally pruned orchards was a substantial rise in the amount of recommended fertilizer application. The more biomass leaving the orchard in the form of almonds, the more nutrients must be replenished to sustain maximum productivity. Between 1973 and 2019, the rate of fertilizer use effectively doubled, from 127 pounds per acre to a whopping 250 pounds per acre,[48] according to university extension economic analyses. In fact, agronomists with support from both public funds and a prominent fertilizer manufacturer argued that 275 pounds per acre (309 kilograms per hectare)[49] was the appropriate application rate. These recommendations rose unabated despite mounting political pressure to address groundwater contamination and the climate impacts of heavy fertilizer use.[50]

A fourth major shift in almond intensification came in the form of irrigating at times when the taps were typically off. In California, harvesting machinery involves three steps: shaking the tree trunk such that nuts fall to the ground, sweeping them into rows, and then hauling them to a processing plant. It used to be that watering was typically cut entirely during that time to keep nuts on the soil surface dry, and growers might not turn it on again until after the winter rains. An extensionist explained that one farm in the San Joaquin Valley had been consistently outproducing its neighbors. When the 380-acre lot was acquired by the corporate giant Paramount Farms (later renamed the Wonderful Company), the same farm manager stayed on and the new owners along with their research collaborators watched closely to see what he was doing differently. They found that the farm manager had turned on the taps as quickly as possible after swiftly removing nuts from the orchard. Researchers then confirmed the significance of this practice, showing that the postharvest period was exceptionally sensitive for almond fruit development in the following season,[51] and extensions agents set out to spread the word.

A suite of yield-increasing techniques in almond production that was fiercely advocated by publicly funded agronomic advisers, and facilitated by a research-oriented corporate grower, all effectively contributed to the rise of K_c. As a university extension seasonal newsletter from 2007 put succinctly, "Why was 100% almond ET estimated to be about 42 inches in the 1980s and 1990s while most growers now apply 48 to 56 inches? Were the earlier numbers wrong? The short answer is NO—the earlier numbers weren't wrong, but many of our earlier assumptions about the performance potential and management of almonds were wrong."[52]

THE SCIENCE AFFORDED BY SCALE

At the center of this new performance potential was the influence of the Wonderful Company under the leadership of a CEO with a geology PhD who was deeply passionate about research. His commitment to finding profit-boosting potential through rigorous study accelerated intensification throughout the state. "When we first got started at Wonderful," a member of the executive team told me, "one of [the] top priorities was to find out how much water almonds actually needed." Previous work on another large operation was guided by general University of California recommendations, but the CEO felt they were not rooted in enough research. He "recognized water as being the most important issue in growing almonds [and] wanted to get right on that. No one had done that research up to that point. Was there a time of year you could shore them? How much water did you need to apply? What was the critical amount, and how did production respond to different amounts of water?" The Wonderful Company's research on postharvest irrigation made a major impact on yield and effectively ratcheted up water use across the industry. "Nowadays, you'll never see a stressed orchard at harvest. People simply will not allow it to happen."

The scale of operations and commitment to scientific study at Wonderful were unparalleled. The company had, at its peak, over one hundred thousand acres (40,469 hectares) of almonds in addition to citrus, pistachios, pomegranates, and more. For researchers, Wonderful was a dream partner, willing to commit the space needed for complex trials with multiple replicates over many years. One irrigation researcher when asked if other farms were asked to participate in a study they had planned with Wonderful said, "You can't. It's too big. It was thirty-eight acres, for five years!" He also saw much less interest from small farms. "A guy with fifty acres, what research is he interested in, if any, versus a guy with fifty thousand acres?" Improvements boosting revenues by a few dollars per acre would make a negligible difference at such a modest scale.

Managing farm trials across many smaller operations requires working with many personalities, another researcher explained, but with Wonderful, "these people are just great cooperators. You go and you've got a trial and all of the sudden the manager's showing up to help you out." A university scientist investigating fertilizer applications in almonds explained, "Doing research on a farm is an imposition, and it requires real management skill by the farmer to ensure that their normal practices don't get overlaid on the top of your experiment." Whereas smaller independent growers might stray from strict research protocols given personal circumstances, responses to market conditions, or simple forgetfulness, Wonderful stuck to the script. "The big growers, particularly if they have a reason and a real interest in it and in the research, are the easiest place to do it, and so that's why it gets often done on big farms rather than small farms." A former Wonderful farm manager explained with pride the lengths that the company would go to for science: "We were willing to go the extra mile to pick up this yield separately" even when "it cost them money for you to slow down," sometimes even ripping out entire healthy trees just to weigh their biomass. Even a fraction of a percent in profitability across such massive acreage meant the research effort would pay for itself. For smaller growers, not so much.

The company not only participated in university-backed experiments with publicly shared results, it also conducted internal studies. The CEO made sure to participate on the research committee of the Almond Board, and if the board didn't agree to fund a study, he might pursue it independently. An extensionist and frequent Wonderful collaborator explained, "When [he] came across a really good researcher, [he] would say, 'Here's what I'm thinking, here's what I'm wondering about. Here's what I'm thinking about. Can we design an experiment to do this?'" This approach was not just for irrigation but all agronomic practices. One scientist described that "Paramount [now Wonderful] has a research conference, just an in-house research conference, where I would be invited, but I couldn't go to the actual conference. I had to stay outside the room. Then when it was my time to talk, I was able to go in and give my talk. Then I had to leave." Even long-term research partners would be kept at arm's length, as the company considered private agronomic research to be a form of trade secret. But farming practices are often visible to passersby in some form or another. When Wonderful began irrigating immediately postharvest, one manager recalled, "we were seeing our neighbors starting to monkey see, monkey do," despite the fact that these same people had at first stopped by to tell them to "turn it off," that it was "madness." When Wonderful chose to stop pruning almost completely, other growers quickly took notice.

The corporate culture of Wonderful's operations also put pressure on mid-level managers to push production higher. Each crop manager or block manager's responsibility was to their boss. The CEO was known to make weekly flights to surveil the company's holdings from the air and follow up with questions about any abnormalities. One former employee working in irrigation described the sentiment among staff. "They better not stress these trees 'cause he's going to see it from above and he will not be happy." An independent farmer might accept, or not notice, occasional deviations from maximum production. But for managers at Wonderful, their job might be on the line.

PLANT STRESS AS UNTAPPED POTENTIAL

An expanding array of technologies have proliferated in recent years to sense plant stress, in addition to crop ET, more continuously and precisely than ever before. Moisture sensors can be buried at varying depths below ground, tree trunks fitted with diameter sensing belts, or drones sent to hover above the tree canopy gathering thermal images. It is theoretically possible for stress-sensing technologies to reduce water use in a given orchard. Known as regulated deficit irrigation, close attention to stress allows for reductions in water to be reserved for the times when it is least likely to impair yields. Plant stress measurements can also signal to growers when to start irrigating in the spring, perhaps delaying the start and allowing the tree to draw from soil wetted by winter rains first. As with drip irrigation, however, plant stress measurements do not inherently help reduce overall use. In fact, the opposite can also be true. According to the people I spoke with who sold these tools and associated software capabilities, growers came to them because they wanted to find the "trouble spots." Agronomists have encouraged this, remarking that many commercial orchards are "underirrigated."[53] Much like drip irrigation was popularized by promising yield increases, precision measurements of plant stress served to seek out with increasing accuracy the moments when additional water could be productively applied. Visualizing plant stress was exciting for almond growers and their advisers because it indicated untapped potential.

As it turns out, detecting excess water is much more challenging than detecting water limitation. There is no clear physiological signal of when an almond tree has been watered beyond its full rate of evapotranspiration. Water not taken up by the plant or evaporating from the soil surface simply remains, slowly flowing toward surface streams or aquifers below. Prolonged saturated soils might eventually cause root rot and other diseases, but it can take years for these problems to emerge. One prominent almond irrigation researcher explained before an audience of farm professionals in 2016, "At some point, too much irrigation should cause problems and reduce yield, not to mention environmental issues, but the 'too much' water point has yet to be determined."[54] The best measure of whether a tree is truly under stress, he explained, is to water it and see if there is a beneficial response. A year later, a different group of researchers found that yields peaked at 49 inches (1,250 millimeters), about 9 percent below the current "crop water requirement" recommendations.[55] More than that would be a waste, but it did not appear to harm the trees. It is unclear whether any growers have taken that recommended reduction to heart.

The materiality of irrigation infrastructure also poses a constraint to water conservation. Many cutting-edge irrigation technologists envision a scenario in which sensors give growers the confidence that they have indeed met their needs in real time and can reduce water applications in specific areas of the orchard. During my research, irrigation was typically managed by adjusting levels for a given block, which could be 50 or even 150 acres in size. A small section of trees might have unique needs, but with current infrastructures, there simply aren't the independently controlled valves, pipes, and tubes to differentiate it from the masses. New variable irrigation technologies promise to allow for irrigation at a finer scale, but this is likely to be at the level of a single acre, not a single tree. Of course, precision does not directly translate into savings. Agricultural extension workers testing the latest generation of precision systems (variable rate irrigation based on plant-level stress measurements) quickly found that even when almond trees were irrigated at full ET$_c$, they were still showing stress![56] The ability to identify trouble spots and then act on them revealed that more water should be applied, even when that was not the study's original aim. While reported as a surprising result, this continues a long-term trend in which precision technology uncovers increasing levels of plant need.

IRRIGATION GOES BEYOND ET

Many California growers have come to rely on irrigation for more than just plant growth, applying additional water usage beyond the already staggering "requirement" of crop ET. The specific knowledge practices centered around measuring and replacing the amount of water flowing through the almond tree have contributed to a secondary problem: salty soil. Irrigation creates a salinity problem as dissolved salts from water and fertilizers accumulate in the soil. Excessive salts make plants work harder to obtain water from the soil and at high levels become toxic. The problem is more pronounced with drip irrigation,[57] where, by design, highly localized applications mean water carrying salts does not flow beyond the root zone. Instead, salts build up at the edge of the saturated zone of a particular tree, forming a bowl underground that eventually compromises tree health.[58]

The only feasible strategy for addressing salt buildup is to saturate the soil at discrete intervals in order to push the salts lower in the soil profile. This additional water, called a leaching fraction, varies widely based on water and soil conditions. In a prime almond-producing region in California's San Joaquin Valley, up to 30 percent additional irrigation can be recommended (nearly 22 inches per acre)[59] to prevent damaging levels of salts, though a 15–20 percent fraction is more common. This is the logic behind the seemingly absurd practice in which growers, particularly those in dry and thus more salt-prone regions, turn on the taps when it's raining. With a boost from natural showers, they can take advantage of a more complete saturation of the soil to push salts deeper. Rather than contribute to tree growth, water applied as a leaching fraction merely prevents tree injury. Unfortunately the desired leaching of salts also promotes the undesired leaching of nitrates, fertilizer residues which contaminate surface and groundwater. While the soil salination problem is as old as irrigated agriculture, the use of precision irrigation based on crop water requirements has magnified its severity.

California almond growers have come to irrigate not only to boost yields and leach salts but also occasionally to ward off the danger of frosts when buds are most sensitive. This works because when air temperatures are below freezing, the irrigation water cools as it reaches the soil and releases heat into the surrounding air. The temperature is only raised by one to two degrees, but this can be enough to prevent devastating damage to sensitive flowers.[60] The exact amount of water released depends on local conditions, but the more saturated the soil, the more effective the frost prevention. Irrigation is not the only strategy, but it is by far the most common. (Some California growers resort to heaters, wind machines, or even hiring a helicopter pilot to hover above the orchard!) Of course the traditional form of frost protection is to plant almonds along hillsides away from cool valley floors, as has been the case in Spain and throughout the Mediterranean, but this is incompatible with California's production maximization modus operandi.

THE CROP WATER REQUIREMENT
ENCODES EXTRACTION

The science and practice of increasingly intensive almond irrigation exemplifies how a drive to maximize production embeds extraction in statements of agronomic fact. The crop water requirement for almonds in the San Joaquin Valley is greater than the rainfall of any location where almond trees can feasibly grow. Almonds need long, hot, dry summers and just enough chill hours each winter to trigger their annual cycle. Commercial production is largely restricted to the Mediterranean, California, Australia, and the Middle East. To plant an almond tree and expect to irrigate at the level of the crop water "requirement" in arid regions is to expect water consistently extracted from elsewhere, either deep underground or farther afield.

The geographical particularity of the almond tree was not incidental to the planting boom of the early 2000s. It was in fact a key contributor. Limited almond-growing land around the globe served as a key selling point attracting growers and investors. Supply would inherently be limited, industry experts explained in conference rooms year after year, and the Almond Board of California had proven its marketing efforts could keep demand strong. The value of land suitable for almonds rose considerably, in no small part based on the expected returns from an intensively irrigated production system. The calculation and escalation of the crop water requirement has both contributed to the almond boom and made its unsustainable water withdrawal rates even more entrenched. Climate change deepens the predicament, cranking up overall evaporative demand via increased heat and aridity.[61] Plants in regions becoming hotter and drier are going to need even more water to keep pace with the productivity expectations of the past.

PRIORITIZING RESILIENCE OVER RETURNS

From the perspective of a Spanish rainfed almond grower, all of these elaborately choreographed calculations are effectively meaningless. Spanish agronomists certainly make use of these metrics, and they regularly engage with their California counterparts, but their measurements circulate only among a small cohort of capital-intensive, irrigated farms jumping in to cash in on the almond boom. Despite the misguided association with water conservation, evapotranspiration and plant stress are the language of extractive operations. Beyond these spaces, they hold little value.

The practices of Spanish rainfed growers are by and large not guided by environmental virtues. The socioecological context of almond landscapes in Mediterranean Spain has simply facilitated a very different way of engaging with plant-water relations. Over the course of the twentieth century, Spain plumbed its rivers with a network of dams and diversions much like, even modeled on, California (see introduction). Intensive precision irrigation followed a similar path, particularly in the lucrative fruit and vegetable industries.[62] Yet almonds were not the crop of capitalists and their state and scientific allies. Quite the contrary, in the eyes of the state, they were an agricultural afterthought of little consequence, with seemingly erratic plantings reflective of an irrational peasantry. The marginality of almonds allowed alternative ways of knowing plant-water relations to root down and persist even when eventually met by resistance from scientific and business communities.

The history of plant-water relations in Spanish almond orchards begins with a very different set of assumptions about the role of almonds in the political ecology of production. Both Californian and Spanish growers in the early twentieth century saw irrigation as wholly unnecessary and even troublesome. Settler orchard capitalists looked at their California almond orchards as vessels for maximizing return on investment in newly acquired land, and by racing to cash in on this new boom crop, they soon faced a constant struggle with overproduction. Peasant producers in Spain, by contrast, experienced no such rush and no such overproduction dilemma. Without a sudden surge in production, there was no fear of a price crash motivating growers to organize and advertise. Without a pool of investment capital flowing into almond orchards, scientists and agrochemical firms showed little interest. Spanish almonds largely remained a marginal part of a diversified farming system. Known to tolerate droughts but also to swing wildly in production from year to year, almonds were typically minor players in sustaining modest rural households. Farms with a significant portion of almonds typically had very challenging terrain with rocky soils or steep slopes and made most of their living off the farm. Admired for being low maintenance, the almond's resilience has been central to its appeal. Almonds in Spain have long been at the economic and agroecological margins.

The absence of irrigation by no means indicates an absence of attention to plant-water relations. While techniques vary widely with local conditions, growers in Spain have used terracing, shallow tillage, and in-situ rootstock propagation to make the most of limited and predictably erratic rainfall. On sloped land, terraces maximize water infiltration by slowing the flow of rainfall and allowing it to percolate down toward the tree roots. Terraced almonds make productive use of steep grades where few other crops can be feasibly maintained. In areas with wider terraces dedicated to grains, almond trees are often planted along the edges such that their roots stabilize the land's structure. Farmers know that moisture stored in the soil risks evaporating from the soil surface, particularly during scorching summers. Breaking up the top few inches of soil into a fine dust with a few shallow tractor passes each season interrupts the conduits of droplets rising toward the surface, serving as a mulch. Not all almond plantings are equally attentive to long-term soil health, and at times government subsidies for new almond plantings have inadvertently incentivized practices with high erosion potential,[63] but soil management has been a primary tool for managing plant-water relations in rainfed regions.

Rainfed growers tend to prioritize resilience over productivity and have continued doing so even when met with sharp criticism by agronomists and almond processors. Many rainfed growers describe the importance of planting almond seeds directly in the ground and then grafting a desired scion onto this rootstock (referred to as *pie franco*), rather than purchasing a tree from a plant nursery. The deep initial taproot, they explain, draws water from far lower beneath the surface than roots that have been kept in shallow pots or been dug up to be transported. Many criticized the heavy promotion of an almond-peach hybrid rootstock known as GF677 by agronomists starting in the 1990s. It grew more quickly to accelerate production and boost yields, but its roots were concentrated near the soil surface. Farmers noticed that when a drought predictably arrived, the tree could not endure as much as the direct-seeded one. Scientists assured farmers that new hybrid rootstocks could tolerate drought just as well. But many farmers weren't convinced that controlled experiments could be relied on to inform the diverse circumstances of working orchards. "These researchers aren't risking much. They work under pressure. . . . For rainfed growers it's much slower. . . . We're more cautious." Many growers also insisted on the use of bitter almonds as a rootstock, believing that they confer drought resilience, as the taproot runs deep and is less palatable to pests. This practice was especially despised by almond processors and other industry figures struggling to ensure buyers that their almonds would be free of the occasional bitter nut.[64] The insistence on choosing rootstocks that prioritize drought resilience over any other feature despite strong objections from agronomic and economic advisers is a testament to its centrality for rainfed growers.

Almonds are planted "where you can't put anything else," as I was told countless times. They occupy a unique agroecological niche. The tree's distinctive capacity to balance its growth with dramatic swings in water availability and extended droughts is its greatest asset. As a compliment to other agricultural and economic activities, optimization is less important than endurance. Agricultural scientists and industry professionals have often spoken of rainfed almonds dismissively, lamenting the failure to increase production and profitability. Yet this marginality and complementarity is precisely what allowed them to persist in balance with their local ecology. For many rural households, almonds provided a way of living within water's limits and maintaining a way of life rooted in the land that otherwise may well have disappeared.

REPRODUCING THE CALIFORNIA RECIPE

When I arrived in Spain in 2017, almonds had hit unprecedented prices, attracting a new set of actors: intensive growers, real estate investors, and a flush of resources from state-backed scientific institutions. The science of almond growing surged along with the influx of capital, as agronomic guidebooks, seminars, and research initiatives offered to bring California-style irrigated production to the Mediterranean. In 2005, just 4.6 percent of almond acreage was irrigated in Spain,[65] and this irrigation was in very modest quantities within predominantly rainfed regions. By 2020, that number hit 19.6 percent,[66] with new acreage overwhelmingly dedicated to intensive California-style plantations stretching across flat expanses fed by dams, canals, and networks of fine black tubing. Good prices prompted a modest uptick in rainfed plantings, hovering at 3–6 percent growth annually, while irrigated plantings soared by more than 28 percent in a single year.

Many growers and agronomists tempered concern about water by explaining that these new irrigated plantings were simply providing "support irrigation" to avoid catastrophic losses or "reduced deficit irrigation" to make the best use of limited resources. Few farmers readily divulge their actual water usage, but the rough estimates I was provided range from nine to nineteen inches (one thousand to two thousand cubic meters per kilometer), merely a third of California's intensive production at the most. In a country with strict water laws like Spain, many farms in almond-growing regions do not have legal rights to much more. Yet the production data tells a different story. By 2021, Spanish intensive almond orchards were almost certainly irrigating on par with their California counterparts. Water usage data may be a private matter, but yield data is public. For almonds, yields track tightly with irrigation. Without California-level water usage, California-level yields would be impossible. Up until 2015, the notion of modest support irrigation appears to have held true. Yields had remained consistent year after year at a level

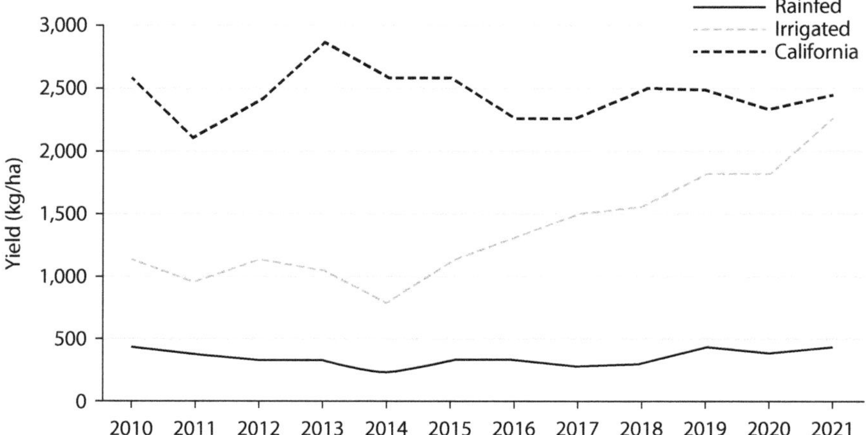

FIGURE 5. Intensification of irrigated production in Spain. Data sourced from annual agricultural statistics published by the USDA (California) and MAPA (Spain). Analysis and figure by author.

that was less than half of California's highly intensive system. But in just five years, irrigated yields surged to nearly meet those of California. Modest irrigation had quickly become a thing of the past. The productivity revolution that took California growers half a century to achieve caught on among Spain's newly irrigated orchards in the blink of an eye. A recipe, once written, is much easier to repeat.

The fact that Spain has a much more stringent water regulation regime than California seems to have had more of an impact on farmers' willingness to disclose their irrigation figures than actual withdrawals. A scathing Greenpeace report in 2017 claimed that, driven by the profitability of intensive agriculture, 70 percent of wells were constructed without authorization and the volume of illegal groundwater abstractions had doubled in the ten years prior.[67] Hydrologic research in Spain has reached the demoralizing conclusion that programs to restrict new wells, pay farmers to reduce withdrawals, or buy water rights have not substantially curbed pumping.[68] Legal scholars refer to this seemingly willful rejection of water law as widespread "hydrologic insubordination."[69] A 2015 study concluded that profitable permanent crops are disproportionately responsible for this trend.[70] Even if on the books almond growers report using less water per hectare than their previous annual crops, like sunflower or maize, a much higher portion of almond water withdrawals are likely to be illegal. The new class of irrigated almond growers in Spain are subject to surface and groundwater laws at the federal and autonomous community level, but the yield data show little sign of reigning in California-level irrigation practices.

Up until the almond boom, very few Spanish agricultural scientists spent their time researching almonds. Research activity focused almost exclusively on breeding new varieties because it was assumed that almond growers would not have the means to obtain irrigation or the fertilizers and pesticides that follow from intensive production. As one agronomist at a regional experiment station explained, "the growing conditions are very limiting. . . . What can we improve from the technical point of view? Very little. For that reason [almonds] are given very little" public research funding. The private sector also took little interest in promoting almond research, given the limited capital of producers to buy their products. Rainfed production is often described as organic by default. "Traditional areas in Spain are marginal . . . but they are very healthy conditions for the crop," another researcher explained. "There's very little incidence of disease." Pesticides are rarely deemed necessary, and most fertilizer is from locally produced manure or plant residues.

One plant breeder who dedicated his career to almonds explained that he was actively discouraged from pursuing almond research given its marginal status. "Among researchers, those who researched fruit trees didn't give any importance to the almond." A colleague of his in Italy, he recalled, was even working on almonds in secret without the knowledge of his boss, knowing he wouldn't approve. Another breeder was driven by a personal passion due to the social significance of almonds. The trees were so widespread across coastal Spain, he explained, that "anything that affected almonds affected countless families." Yet they received few resources for research because almost no one relied on almonds as their primary income. Spain's most lucrative crops—olives, wine grapes, citrus fruits, and greenhouse horticulture—dominated agricultural institutes. While in the past almonds were "marginal," a surge of capital meant they suddenly attained the status of a "frutal" (fruit tree).

This new identity as a frutal signaled that the almond had now been drawn into the spheres of agribusiness, a ripe new market for plant nurseries, irrigation equipment manufacturers, and vendors of fertility and pest control products. State-funded agricultural institutes ramped up their involvement in all things almond, engaging not only in research activities but also public workshops and farm tours. Experts from California universities were invited to give talks about the keys to success, and seasoned California almond crop advisers were hired by wealthy landowners to advise their new operations. As a PhD student in agronomy preparing to defend her thesis on almond production explained candidly, now that almonds were being planted in fertile soils with intensive practices, they were worth studying.

New corporate farms were sprouting up rapidly to jump on the almond bandwagon and attracting young agronomists trained in intensive fruiticulture to fill their ranks. Names like Ocean Almond and Agro Water Almond featured English and aquatic terminology to assert their "modern" outlook, one employee explained. These orchards were investments made by real estate, hospitality, and construction firms seeking a place to sink their capital while those core industries were in decline. Buying or renting land went quickly given their connections as developers, I was told, often with unusually large parcels of 600 hectares (1,483 acres) or more. Many of these investor-owned operations were experimenting with some parcels planted in even higher planting densities than that of California. These super-high-density or hedge-style plantings were a relatively untested concept for almonds that had been borrowed from the intensive olive industry,[71] and many agronomists remained skeptical of the long-term benefits. The primary justification I was given in interviews with farm managers was the hope of boosting production in the orchard's early years to capitalize on boom-time prices that were unlikely to last. Tree nursery businesses were also inclined to promote high densities, which meant higher sales. The latest in precision technologies promised to monitor orchards in real time, with rapid response to avoid losses. Even a salaried crop adviser bringing these new plantations into being was awestruck by the application of intensive irrigation to almonds, confessing that, given where he was raised, "for me, almonds will always be a rainfed tree." The rules of plant-water relations were being rewritten before his eyes.

Not all the intensive plantings were run by investors. Much of the newly planted almond acreage was also comprised of locally owned irrigated operations seeking an escape from the sinking prices of grains and sunflower oil. A farmer in Zaragoza, Spain, explained that he and his partners who farm 667 acres (270 hectares) of olives, peaches, persimmons, and now 40 acres (16 hectares) of almonds were the "four crazies with irrigated almonds." Almonds, he explained, had always been considered "for the steep slopes, historically, not like a crop." Now their neighbors were no longer calling them crazy and instead were racing to plant their own irrigated almond land. Another young man in Zaragoza with about 494 acres (200 hectares) of irrigated grains decided to switch a third of his operation to almonds, seeing the strong prices, and was devouring all the agronomic guidance from California he could find to know what works best. With some basic English and the help of online translation tools, he could easily access recent publications and adapt them to his farm. An agronomist farmer I met with a career in intensive citrus and stone fruit production explained how odd it was to reorient toward almonds and realize that "this is a tree that needs much more," given that "almond trees have always been at the margins, where you couldn't plant grains and there wasn't water for anything else." What had just a few years prior seemed absurd was quickly becoming the new standard.

SEEING STRESSES OR STRENGTHS

How much water almonds "need" is less a function of flows through plant tissues than flows of capital. The upward creep of the crop water requirement demonstrates how a seemingly stable agronomic fact is in reality a reflection of increasingly intensive practices designed to maximize returns. Precision irrigation management, whether via evapotranspiration calculations or in direct response to stress measurements, has long been a yield maximization strategy. The science of intensive irrigation was facilitated by state research institutions and infrastructures but also accelerated by science-driven corporate management. When the sky-high prices of a mammoth marketing investment prompted a planting boom, the agronomic assumptions built into irrigation science leaped quickly across borders.

Crop evapotranspiration, while a seemingly neutral or even beneficial precision measurement of plant-water relations, codifies extractive logics. By requiring more water than falls as rain anywhere that almonds can grow, the calculation enshrines a willful ignorance of local ecologies. In the realm of water exploitation, which neither California nor Spain have successfully reigned in despite intensifying political effort, such a narrow view is inexcusably naive. The absurdity is even more stark given a plant uniquely capable of making the most of drought. But it is not the only way almond growers can work with water. Rainfed growers relying on almonds to fill a niche in a more complex landscape and livelihood prioritize resilience. Their practices are rooted in pragmatism. Capitalist agronomic knowledge strives to identify every possible deficit or stress but remains unable to see a plant's strengths. Given the right social conditions, a plant notorious for its thirst can just as easily be championed for its thrift.

One might counter my examination of agronomic calculations here by arguing that irrigation science simply informs growers about productive possibilities and that the real problem is cheap and insufficiently regulated water. These are important points to consider. The almond boom, as it turns out, demonstrates a scenario in which the effectiveness of water pricing and law are seriously called into question. In California, where water markets are increasingly popular, price spikes in which water for some users rose to forty times the typical price (from $55 to $2,200 per acre-foot)[72] did not appear to significantly alter almond irrigation practices. There were two-year waiting lists to drill million-dollar wells. Anticipated profits justified such expenses. In some cases, growers with limited or highly priced water removed mature orchards and planted new seedlings, which demanded less water in the short term while banking on more abundant flows in the future. This is not to say that the drought didn't sting for farmers. Expensive water ate into expected profits, and many smaller growers couldn't afford the costly measures of their larger counterparts. Some with limited supplies let other annual crops go fallow to ensure their almonds were well watered. But when it came to almond yields per acre, effectively a proxy for irrigation, there was only a slight dip, which quickly rebounded.[73] The drought put a damper on the almond boom, but acreage is still expanding. As agricultural economists and industry members note, almonds make economic sense because when water is expensive, they generate a high return on irrigation, or dollar per drop. Making water more expensive is undoubtedly a useful tactic for reigning in usage. But given the almond industry's impressive ability to grow demand and keep almond prices high, the water price that could significantly cut back almond irrigation behavior might be astronomical. And of course, there are equity concerns about the consequences of simply sending water to the highest bidder.[74]

Pointing to the cost of water also takes for granted the ability of the state to adequately monitor and enforce water regulations. In Spain, where laws restricting both surface and groundwater use have been firmly in place for decades, illegal activity is rampant.[75] Almond yield data indicates that Spain's irrigated growers have effectively hit California levels of water use without water restrictions seeming to get in their way. Effective legal enforcement of groundwater policy is a global predicament with painfully few success stories, and it is virtually unheard of in regions dominated by highly profitable irrigated agriculture.[76] Keeping tabs on wells, which are widely dispersed across private properties, proves challenging, and governments often lack the political will to counter their powerful farmer constituents. California has long struggled to assess the impacts of illegal water withdrawals,[77] let alone prevent them, and there are strong indications that the state's nascent groundwater enforcement institutions have been captured by powerful grower groups.[78] Water policies are crucial, but it is not so simple to say policy alone, whether market-based or otherwise, would fully address the problem.

The notion that agronomists merely inform and policy or economics influence actual practices is a convenient illusion in which science appears isolated from, rather than constituted within, social worlds. There is little doubt that if the crop water requirement for almonds did not exist, or had not been meticulously studied and augmented over time in association with other intensification research, fewer growers both in California and Spain would water as much as they do now. For these industry professionals, the goal is to help growers do well, and if improving the recipe means adding more water, they will dutifully convey that data. Their work has been fundamentally shaped by capital flows—from agribusiness partners donating materials, to university funding priorities, to the ease of cooperation with corporate farms—as well as their own desires to feel valued by their community, yet the outcomes become purified through the ostensibly value-neutral language of science. Researchers find themselves in a complex position, keenly aware of environmental harms and yet actively enabling agricultural practices they know cannot last.

Agronomic knowledge about what a crop "requires" has long masqueraded as intrinsic, placeless, or value-free. Yet the details regarding who participates in agricultural knowledge dissemination, under what conditions, and with what resources and priorities have shaped the plant-water relations that prevail. A relentless focus on efficiency and precision in the hope that a profit-boosting tactic will conveniently deliver ecological integrity is doomed to fall short. The lessons to be learned from rainfed growers are not a set of practices but a set of alternative values: longevity, thrift, and the rootedness not only of trees but rural communities. What might an agricultural knowledge infrastructure look like that privileges ecological and social stability, resourcefulness, and a holistic integration with a specific place? What if the agricultural margins, where low-input and ecologically suited production persists, were a priority for public institutions and encouraged to thrive rather than assumed to vanish? What if agronomic science that excluded its ecological context was resoundingly rejected as incomplete? Admitting the politics of producing agricultural "facts" is not critique for the sake of critique; it affirms the possibility that things could be otherwise, and that it will take a different type of politics to get there.

3

Symbiosis

Producing Pollinator Dependence

There's a steady hum in the air as dawn casts its light over the pale pinkish blossoms of an almond orchard in California's Central Valley. White wooden boxes marked with spray paint and stencil, each containing a colony of many thousands of honeybees, are poised along the roadside. With any luck, the February skies are clear and the cool evening air quickly warms, beckoning the bees to take flight. Each flower they explore is an almond yet to be, awaiting pollen brushed from the bee's body to trigger transformation. Their hive came hundreds, likely thousands, of miles for this delicate task. So did nearly every commercially managed colony across the continental United States. More than two million hives ordinarily spread from the Pacific to Atlantic coasts pack into just over two thousand square miles (5,180 square kilometers) of orchards, an area roughly the size of Delaware or twice the size of Luxembourg. The density of honeybees buzzing about such a narrow strip of earth is unparalleled. This is the greatest managed pollination event in history.

Every year beekeepers from across the country, even the farthest stretches of Florida and Maine, load their hives onto trucks bound for California. While many families have been beekeeping for generations, the industry is no longer a mom-and-pop business. Commercial beekeepers manage tens of thousands of hives, often hiring workers with origins in Mexico and Central America for the most hands-on tasks. Experiences of migration are layered upon one another. Bee workers drive long hours at night when the bees are not interested in leaving their boxes to forage, navigate time-consuming state inspection checkpoints, and finally unload the buzzing cargo, praying the heavy treads of forklift tires can make tracks through the slippery spring mud.

While the hives may remain in California for only a few weeks, they have been subject to a year of increasingly intensive preparations. This pilgrimage is the

FIGURE 6. Commercially managed honeybee hives placed within an intensive almond orchard in Fresno County, California. Photo by author.

FIGURE 7. Rainfed almonds growing along terraces between wine grapes and forested areas in Valencia, Spain. Photo by author.

lifeblood of the contemporary American beekeeping industry, and yet it is also one of its most formidable challenges. With a bit of luck, these bees will manage to dodge the pesticide sprays of neighboring fields, avoid picking up too many diseases and parasites from nearby hives, and leave before a monocultural flower feast quickly turns to famine. The labors of keeping bees healthy enough to work the almond orchards despite a barrage of threats to their well-being can be arduous and stressful. But for many beekeepers these days, opting to stay home effectively means going out of business. Almonds need bees, and beekeepers need almonds. The scale of this economic and ecological symbiosis sparks a combination of awe and anxiety. It is highly profitable and highly precarious.

In Spain, the picture is quite different. Pollinators suffer the same pressures, and honeybee hives are trucked to and fro across the country to seek out flowering forage and occasionally provide pollination services, but not for almonds (at least not until quite recently). In my interviews, nearly all rainfed growers said that almond trees simply did not need honeybees. Among a new crop of intensive irrigated plantings, the details of pollination service provisioning were just starting to be worked out. There are many reasons for this, all stemming from the socioecological marginality of almonds, as I will elaborate. In doing so, I interrogate what exactly is meant by the claim that a crop needs bees and the conditions through which this need is produced.

What does it mean for a crop to require insect pollination? For decades scientists around the globe have been ringing the alarm bell over pollinator decline and its potential impacts on agricultural productivity. Farming practices, such as pesticide use and managed hive migration, are merely a few factors in the "death by a thousand cuts" devastating insect populations.[1] The varroa mite, an invasive parasite officially deemed *Varroa destructor* for its devastating impacts on European honeybees around the world, feeds on bee flesh and brood while acting as a vector for a host of debilitating viruses.[2] The globalization of disease has exposed honeybees to an expanding range of fungal and viral pathogens.[3] Land use change such as agricultural intensification and urban and suburban development have made it harder for bees to find the floral foods they need to sustain healthy hives that can fend off disease.[4] Common pesticides and their additives impair bees' immune response, leaving them more susceptible to other threats.[5] Ironically, as the combined pressures on honeybee health grow, so too have the high-value agricultural industries that depend on them.[6] Given the oft-cited statistic that nearly 75 percent of crops depend on pollination,[7] the situation appears bleak.

Upon closer inspection, pollinator dependence is far less black and white, and its significance for agriculture far less existential, than such a figure suggests. Insect pollination is entirely irrelevant for the majority of staple foods, including wheat, rice, maize, potatoes, yams, and cassava. In terms of volume, roughly 35 percent of total global crop production relies to some extent on pollination.[8] But even that is a bit of an overstatement. This simple figure, frequently repeated by the US Department of Agriculture,[9] overstates the true impact as understood by the scientists who published the original analysis. Dependence on pollination exists across a spectrum, and these figures represent the broadest possible characterization. Most crops show increased yields or enhanced quality with insect pollination, but for most it is not an absolute necessity for reproduction. In fact, one astonishing study estimated that if all animal pollination were to be eliminated, global agricultural production would drop only 3–8 percent.[10] This reflects the fact that crops highly dependent on pollinators are relatively few (around 10 percent) and are produced in low volumes (just 2 percent of global production). Some scientists have thus questioned if pollinator declines are really a crisis at all when it comes to food production.[11] Pollinator populations and biodiversity are being devastated, there's no doubt, but using agriculture to rally public concern is a bit misleading. A world without honeybees would make many desirable foods like chocolate and melons quite scarce and certainly would reduce important cultural and dietary diversity, but it would not provoke a global famine.

Perhaps a more candid representation of today's pollinator "crisis" for agriculture is one of profitability, both for the beekeeper and the farmer. The overproduction symptomatic of industrial agriculture has burdened each with a deep dependency for the other that for many crops need not have ever emerged. It is an industrial symbiosis,[12] not an evolutionary relationship but a socially produced one—an arranged marriage that may or may not work out in the long run. The story of how almonds and commercial honeybees became intertwined in California, and remained largely independent in Spain, illustrates the political economic drivers behind the "need" for honeybees. Bees, like irrigation, are an input on the technological treadmill for farmers. But they are not only that. Mapping the almond-bee relation across space and time reveals the politics of knowledge surrounding the often taken for granted notion of pollinator dependence.

CALIFORNIA BEEKEEPERS COURTED FARMERS FIRST

Unlike other inputs, like water or agrichemicals, honeybees are an agricultural industry in and of themselves. Beekeepers talk of their hives much like ranchers talk about their herds, seeking out good "pasture" and ensuring high quality "forage." In California it was beekeepers in the early twentieth century who first began seeking out farms with suitable crops (and other rural land with wildflowers) to boost their honey production and increase the scale of their operations.[13] The automobile transformed beekeeping from a predominantly sedentary practice, limited by the availability of flowering plants within range of the hive, to one that could profit from mobility. A natural cap on the number of hives one could manage was lifted. Migration opened a frontier not only of space but of time; colonies could be moved to maximize their proximity to flowers blooming at different times, following the various blooms cycles of crops or accessing wildflowers at different altitudes. Fossil fuel transportation in many ways boosted honey production per beekeeper the way synthetic fertilizer (also produced by fossil fuel combustion) boosted yields per acre.

The distinction between beekeeper and farmer is actually quite a recent phenomenon. Keeping bees had long been a side business for orchardists. It wasn't until cars and trucks came along in the early twentieth century that beekeeping became a full-time, specialized, commercial enterprise. The affordable, flexible transportation of the automobile meant beekeepers were no longer tied to the flowers on a specific plot of land. They were now mobile, able to drum up more honey by pursuing new pastures. Keeping bees on monoculture farmland year-round makes little sense as the single crop cannot provide flowers all season long. But for a mobile beekeeper, a monoculture in bloom is like an all-you-can-eat buffet, and when the food runs out you simply move on to the next. The separation of beekeeping from farming was also marked by emerging health challenges of European honeybees in North America. In 1925, a California researcher explained that, due to the time required to manage bee diseases, "honey production is no longer a profitable side issue for the fruit grower or general farmer."[14] Fortunately, he followed, "California has perhaps more 'one crop farmers' than any other state," and "this affords an excellent opportunity to the beekeeper who wishes to rent his bees."[15] Beekeepers needed to move among farms before the farmers voiced a strong desire for bees.

The concept of honeybee pollination as a service or "requirement" for agriculture was driven less by the biological needs of plants than by the economics of fossil-fueled agricultural specialization. Automobiles and an array of geographically and temporally distinct flowering monocultures created a sort of spatial fix for the problem of scaling up beekeeping in order to sustain its profits.[16] For beekeepers, migration was a way to intensify, incurring more expenses but yielding more honey. They also piggybacked on the intensification of agricultural landscapes, particularly the expanding use of irrigation, which enabled much higher populations of bees in dry years than would have been supported by rainfed forage.[17]

For early almond growers in California, honeybees were considered a luxury but by no means a necessity. Like most other farmers at the time, they counted on wild pollinators.[18] Reports of almond growing practices from the early twentieth century make no mention of orchardists interacting with beekeepers or keeping their own beehives.[19] An official agricultural extension instructional book from 1925 shows keen awareness of the need for insect-mediated cross-pollination, but the sufficiency of native pollinators from the surrounding landscape was never called into question. The issue of importance for farmers was to plant the correct varieties such that their bloom timing would overlap and allow local bees to do their work.[20] Some university extension agents pointed out at the time that "it will pay the grower to keep bees,"[21] noting an uptick in yields for the most enterprising growers willing to make the effort. A 1927 study noted that bringing bees into the orchards during bloom was starting to catch on among enlightened growers, but paying to rent them was quite unusual.[22]

By midcentury, a strong mutualistic relationship had developed in which beekeepers from Southern California would bring hives to almond orchards farther north. Often no money was exchanged, or a minimal fee was paid to cover the costs of transportation.[23] Then the dependency deepened on both sides. Beekeepers made their living from honey, and it was becoming more difficult for them to find pasture as the Southern California landscape urbanized and farms specialized.[24] Almonds are not particularly good for honey production directly. In fact, honey from almond blossoms is typically considered too bitter to be salable,[25] but they bloom in winter when flowers are few and hives are at their weakest. The time spent in almond orchards allows bees to strengthen their ranks by feeding the bitter honey to their young before being relocated to oranges or other high-quality honey forage.[26] Bringing hives to the almond bloom early in the year would mean a more bountiful honey flow for the rest of the season.

California almond growers at first considered honeybees to be a convenient yield boost incurred at minimal or no cost, but much like other inputs on the technological treadmill, they quickly got hooked. An entrepreneurial edge became a dependency. This addiction was intertwined with others. As farmers' yields swelled with irrigation and fertilizer applications, the density of flowers (which then become almonds) rapidly multiplied. In the 1920s, yields averaged 211 pounds per acre. By the 1950s, that figure had more than doubled, to 561 pounds per acre, given additional water.[27] In rainfed orchards with fewer flowers, wild pollinators sufficed, but in intensively managed orchards, they were becoming a limit to reaching the orchard's full productive potential. Across the United States, university extension researchers played a major role in convincing farmers that renting bees would be profitable.[28] By the late 1950s, studies showed that almonds in particular saw a major boost from bees because they drop very few fertilized flowers, unlike most other tree crops, meaning growers could aim for 100 percent pollination to obtain maximum yield.[29] In addition, almonds tend to fetch higher prices when smaller in size, and a higher successful fruit set makes for smaller nuts.[30] Researchers argued for a high density of bees, either more hives or stronger colonies, to achieve maximum pollination given the cold temperatures during the February almond bloom, when honeybees actively forage for only a few hours at midday.[31] But it took a good deal of convincing for growers to start spending on something that had historically been free. It was not until 1963 that the cost of bee rentals, at two dollars per hive, was reported as standard practice by university agricultural economists.[32]

Farmers were pushed to pay for bees not only because their own operations had become more intensive but also because native pollinators were no longer as prevalent or reliable. Insecticides were taking a toll, and the popularization of "clean cultivation," or tillage between rows, eliminated crucial wildflowers and bee nesting sites.[33] University experts began recommending two hives per acre, evenly distributed, to ensure no yield losses could be attributed to lack of pollination. By 1973 the almond industry outgrew beekeepers within the state and began drawing bees from across California's borders.[34] Most out-of-state beekeepers at the time were eager to take advantage of the early blooms as the intensification of rural landscapes across the country was making it harder for them to find suitable locations to park their hives.[35] Suburban sprawl and the increasingly frequent rain of pesticides across American farmlands progressively narrowed safe spaces for honeybees to thrive.[36] Then in the 1990s, a plague by the name of *Varroa destructor*, a mite that impairs bee bodies and acts as a vector for numerous other ailments, began wreaking havoc on American beekeepers. The costs of battling varroa made the income from pollination services that much more important to balance the books. Strong almond prices meant growers were willing to pay more and more to bring bees to the bloom.

These trends continued for decades, almond orchards steadily expanding and beekeepers increasingly struggling, until eventually the beekeeping industry turned upside down. In the early decades of the twenty-first century commercial beekeepers became providers of pollination services first and honey second.[37] Almonds were booming just as beekeeping was buckling. In 2017, almonds accounted for 80 percent of all pollination service payments.[38] One established California beekeeper explained in our interview,

> Say twenty years ago, if I had ten thousand hives, my best year, twenty years, I could have made around 1,500 barrels of honey in one year. And right now, the same amount of hives I have right now, because of foraging and problems and not enough food for the bees to collect the nectar, I am averaging anywhere between 250 to 350 barrels a year. . . . And that's not just me, it's everybody, and I have really good locations because I've been doing it for so long. A lot of us guys that have been in the business a long time have some of the better locations. Then you've got these other newer beekeepers that, like I said, they don't even try to make honey, because there is nowhere to make honey.

For some beekeepers traveling thousands of miles from Georgia or Florida, the switch happened practically overnight. "For the thirty, twenty-nine years of my bee business, 100 percent came from honey production. The last four years, probably 60 percent to 70 percent comes from California almond pollination. . . . Any honey we make and everything is just more or less bonuses really for us." Migratory pollination is no longer a means to an end; it *is* the end. US beekeepers are fully almond-dependent.

Meanwhile, almond growers had become hooked on honeybee pollination and willing to pay previously unimaginable sums to get it. In 2006, pollination service fees nearly hit two hundred dollars per hive, or four hundred dollars per acre. Growers were starting to top two thousand pounds per acre on a regular basis, churning out more almonds from the same plot of land by increasing planting densities, water applications, fertilizers, and pest control measures (see chapter 2). There was simply no way that wild pollinators, even if their populations were protected from industrialization's ills, could sustain such yields on their own. While almond growers in the early twentieth century wouldn't have dreamed of paying to truck honeybees to their orchards, almonds had now become the quintessential pollinator-dependent crop. This dependency is not an inherent botanical characteristic, it has been produced by the intensification of almond orchards and the broader industrialization of landscapes made inhospitable to pollinators.

The Price of Dependency

The presumed pollinator dependence of almonds has serious consequences. It has effectively remade beekeeping in industrial agriculture's image, relying on a mounting suite of toxic agrichemicals and resource-intensive feedstocks along with a heightened risk of pesticide exposures. As of 2023, more than 80 percent of commercial honeybees across the continental United States, some estimate 90 percent or more, pollinate almonds each year.[39] For beekeepers, this migration is more economically essential than ever as honey prices have cratered (due to cheap, sometimes adulterated, imports)[40] and good forage has become scarce. Yet it also entails significant costs. Bringing bees to almonds compounds the mounting stresses from pesticides, drought, disease, and poor nutrition. The number of commercial hives in the United States remains remarkably high—some would argue too high[41]—but they are effectively on life support. Beekeepers used to expect they would lose 5–10 percent of their hives over the winter for a variety of reasons. Now 30–40 percent losses are expected, with more than one beekeeper I spoke with saying he'd seen as high as an 80 percent loss in a single year. Yet with so few landscapes providing adequate nourishment, beekeepers are far more worried about there being too many bees than too few. They use a range of techniques to split hives, build the bees up, and address their ails. Supplying strong hives for almond growers is getting more laborious and costly, but pollination prices more than make up for it. Financially speaking, times have never been better. Ecologically speaking, beekeeping has never been worse.[42]

Beekeeping in the continental United States now, quite literally, revolves around the California almond bloom. Yet this orbit, drawing so many bees from across the country to one narrow strip of land, makes the struggle to maintain bee health that much harder. Bees pay no attention to property lines and often fly up to two miles from the hive, putting them in range of hundreds of other hives from distant locales. A Florida beekeeper who had just begun bringing his bees to almonds described California's Central Valley as "the biggest cesspool there is. . . . You literally have bees from all over the United States going to one place, one time a year, all at the same time. Well, as a researcher, it doesn't scare the hell out of you? Because it scares the hell out of me." In addition to the sheer volume of potential disease exposures, not all beekeepers are equally meticulous about the costly process of checking their bees for disease and treating for mites. A long-time California beekeeper and researcher described how "the diseases, the sharing of parasites and pathogens, has just been huge because you get put in cheek to jowl next to every dirty beekeeper in the country." Scientific assessments tend to confirm beekeepers' fears, though they are less assertive, given the challenge of controlling experiments for all potential variables.[43] Among the beekeepers I spoke with, there was no doubt that the almond bloom put their bees at greater risk of disease.

Catering to the almond industry also puts pressure on beekeepers to manage their hives differently. The almond bloom occurs at the hive's weakest point in its annual cycle, and yet growers increasingly demand stronger hives holding more bees. During the soaring almond prices of the 2010s, many growers began paying a substantial bonus for hives containing more than eight frames and rejecting hives with fewer.[44] This level of scrutiny was new, and for many beekeepers it signaled a qualitative change in their relationship with growers. What was once a mutual understanding among two kinds of farmers, that some years are better than others, became a performance-based transaction. To ensure full payment, beekeepers would need to do everything in their power to ensure full hives, regardless of the fact that so much about hive health remained outside their control. One thing they could control, to an extent, was food.

Beekeepers have since, at least the mid-twentieth century, compensated for limited forage on occasion by feeding their bees sugar syrup (typically corn-based), imported pollen, or pollen substitutes (typically soy-based). This used to be a last-ditch measure to cope with unpredictably bad years, like a prolonged drought, and it can never match the quality of natural forage. But given pressures on the industry, the quantities and frequencies of these additions have reached unprecedented levels. Semitankers full of corn syrup regularly deliver to bee yards as if they were petrol deliveries to a gas station. A Florida-based beekeeper estimated, "In my five thousand hives, I'm putting in probably forty thousand pounds of pollen substitute every year, if not more, then the corn syrup. Then there's the corn syrup's running thirty-six cents a pound. Every time I have a semi [truck] come in here, I'm probably feeding two tanker loads a month to keep them big enough." Some beekeepers described an unease with pushing the bees this hard. "Us capitalists, us scientists and geniuses, throwing more food at them with some patty-cakes here, so they got to keep laying in till November, and they're like, 'No, we don't want to. Why you doing this to us?' We need you strong. We need you healthy." The bees have fewer flowers to feed on, are getting sick more frequently, are subject to competition and contagion from living in high densities, and on top of all that they are being pushed to maintain large populations well beyond the seasonal cycles that govern bee behavior. For the hives to endure harsh conditions year-round and provide abnormally bountiful bees for almond pollination come February, an enormous amount of inputs are required that would have been an extreme measure just a single generation ago. Beehives that used to provision entirely for themselves are now on a steady diet of industrial corn and soy.

Disease treatments have also become more frequent and intensive, putting beekeepers on their own pesticide treadmill. A California beekeeper who got into the business working for his father-in-law described, "We didn't start [treating for mites] until I think in 1990, when we started. And then it was twice a year maybe, and then that product quit working. Then we had another product that worked for a few years, and so that's the way it went. . . . Then it just kept getting more and more, or we had to treat more and more all the time." Treating every ten days would now not be considered uncommon, and resistance is a growing concern.[45] When restrictions have been placed on the miticide of choice due to its toxicity, beekeepers have even resorted to extralegal measures. "I hate to say it, we have to buy it on the black market. You meet in a little parking lot in a dark corner around here. You'd think a drug deal is going down, and they're not. It's a beekeeper buying Amitraz." The miticide treatments carry spillover effects. "The remedies that we use to kill the mites is really hard on bees. It can even shut the queens down from laying for ten days, and it messes up the pheromones set in the hive. Just gets to where it's really tough to keep these mites at bay." While treatments would occur regardless of almond pollination contracts, the demands of almond growers are top of mind when making these investments in time and resources. "Because of the nature of pollination and trying to have all these bees and have them all strong, there's tons of input." Immense resources—sourced from corn, soy, and agrichemicals—which would have seemed outrageous just a decade or two prior, have become standard practice for commercial beekeepers serving the almond industry. As one beekeeper summarized, "the cost of getting bees through the winter has gone up exponentially."

Placing bees in farms for pollination, rather than cattle pasture or wildflowers for honey production, also means heightened risk of pesticide exposure. The Almond Board of California has taken great pains to instruct growers on best management practices to avoid harming bees, such as spraying at night, when honeybees don't fly, and using less harmful products. Many have taken these recommendations to heart, but beekeepers know that ultimately ensuring a profitable crop is their top priority. "Ninety-five percent [of almond growers] are well aware of what is best for bees. They don't want to hurt the bees they are paying so much for. But they have to protect their crop. They are going to spray. We know that's going in. Any beekeeper that thinks he's not going to get sprayed isn't paying attention," explained one of California's largest beekeepers. Even if the grower has been well-informed, the decision on how, when, and what to spray often comes from a hired pest control adviser or contracted pesticide applicator who is on a tight schedule and not as directly impacted by bee health. Spraying at night is a hassle. Beekeepers are also increasingly concerned not only with chemicals sprayed in the air but also those mixed with irrigation water. Drip irrigation has facilitated the precision distribution of pesticides for soil pathogens, referred to as chemigation. Unfortunately, bees stop to drink at little pools of water on the surface, slurping up whatever this chemical cocktail contains. The most conscientious almond farmer following all the recommended best practices is still unlikely to eliminate exposures entirely.

According to beekeepers I spoke with, the biggest problem is not actually almond farms but their neighbors. Honeybees flock to alfalfa flowers, which just so happen to receive insecticide sprays around the same time as the almond bloom. "It's usually not the almond guys," a beekeeper explained about losing his hives to pesticide sprays. "It's the neighbor doing something to the alfalfa was the biggest culprit." In California, unlike most other states, beekeepers who register the locations of their bees receive notice forty-eight hours before any potentially harmful sprays so that they have the option to move bees, but it's a logistic nightmare. "It's so bad, the spraying. You move them to another location, and then the next day you get a call from that new area, 'Hey, we're going to spray in a few days.' My record is nine loads I moved at nine different times within three weeks, just to avoid them from spray. By the time I got done moving in the nighttime, they were just as screwed up as if I had left them where they were the first time and let them get sprayed a little bit." Yet the attitude of beekeepers is one of acceptance rooted in economic dependence. "I've had bee losses on almonds before, but we don't usually kick and scream too much because they usually pay so much money."

Beekeepers and almond growers have grown so mutually dependent that it has created a positive feedback loop. Beekeepers have turned to pollination to compensate for low honey production and low honey prices, yet catering to the almond industry is reorienting the ways bees are managed to privilege pollination over honey. "It's getting to the point now where beekeepers nowadays are so worried about keeping their numbers up for their pollination contracts for almonds that they're not worried about putting bees in honey flows." Losing interest in honey production exacerbates the reliance on sugar syrups, as beekeepers have less incentive to seek out high-quality flower forage. "To go to the pollination is a year's worth of work, and you really sacrifice a lot of honey crops just to stay in the pollination with the almonds and take the risk that you have to take." These tradeoffs are reflected in the numbers reported by the US Department of Agriculture: more hives every year with less honey per hive.[46] These hives are pollination providers first and honey producers second. Beekeepers, who a century before sought out almond orchards to crank up honey production, now see almond pollination as their last lifeline. "The whole bee business, the whole industry, revolves around almonds. The whole industry! If that industry collapses, the bee industry collapses with it."

Intensive almond production and commercial beekeeping have become profoundly codependent, but this is a political economic, not biophysical, fact. It is a symptom of agricultural industrialization on both sides. Beekeepers intensified honey production through fossil-fueled transportation. Monoculture farms of increasing scale along with urbanization demanded more distant movements to find adequate honeybee forage. Meanwhile, irrigation and associated agronomic intensification in California pumped more almond blossoms from every acre. Public researchers provided growers scientific assurance that importing honeybees would pay and that almonds in particular could push yields through maximizing pollination. As managed honeybees in the 1990s progressively succumbed to the costly pressures of pesticides, poor nutrition, and disease, paired with low prices, honey production slowed while pollination took priority. Serving the almond industry has become both a blessing and a curse for beekeepers, a lucrative escape from the honey game but also a treadmill of feeds and treatments to prop up hive health amid intensified stressors. The accumulated short-term remedies for the ills of intensification in both industries appear like a house of cards. It stands because the almond boom has made record profits possible on both ends. But the changing of economic winds could send the whole delicate arrangement crashing down.

SPAIN'S SYMBIOSIS AT THE SIDELINES

The California almond industry's cross-continental pollinator pilgrimage appears even more astounding when seen from the vantage point of rural Spain. Almond growers in rainfed regions view managed honeybees as entirely optional. They could be nice to have but nothing close to a necessity. Nearly every rainfed grower I interviewed said that almonds do not need honeybees. Some farmers might host the hives of a beekeeping friend of the family, but at no cost and without any effort to seek them out. In fact, at the end of the year, these landowners are typically gifted a few jugs of honey as thanks for their hospitality. When I visited in 2018, it was rumored that larger wealthier farms might rent hives for pollination, given a spectacular jump in almond prices, but the practice was notable for its novelty. By and large, rainfed almond growers simply do not concern themselves with managing pollinators. They trust nature will do its work. At first glance, some might characterize this as naive or simply a stage along a trajectory toward a more "mature" industry like that of California. There is a much more dynamic story here. Marginality has been core to the way farmers and scientists know and care for almonds in Spain, rooting pollination knowledge in values of resilience and thrift. As a result, the social relations, geography, and science surrounding almond pollination have functioned to encourage independence from intensive beekeeping rather than to foster dependence.

It's worth mentioning that commercial beekeepers in Spain share a great deal in common with their counterparts in the United States. Spanish honey production throughout the twentieth century became increasingly intensified through migration between regions to access forage.[47] The varroa mite and associated diseases have devastated hives, exacerbating the threats posed by pesticides and poor nutrition. Miticides are increasingly essential to keep hives productive, and resistance is likely to make control more challenging.[48] Much like in the United States, good locations for bees are hard to come by due to expanding urban developments, widespread pesticide use, and the industrialization of the countryside, a dynamic worsened by droughts that limit wildflower growth. To cope with these conditions, Spanish beekeepers are starting to feed their bees sugar syrups and pollen substitutes,[49] albeit at much lower levels than those drawn into the California almond orbit. Spain has more honeybee hives than any other country in the European Union, reporting more colonies than the entirety of the contiguous United States,[50] despite a land area roughly one-fifteenth the size.[51] Yet in a 2016 survey of beekeeping across Spain's Mediterranean regions, only 15 percent of beekeepers reported performing any pollination services for farmers.[52] When they did, it was primarily for intensive greenhouse-grown vegetables. Honey is the name of the game, along with more minor revenues from wax and other coproducts. Only since the price of almonds shot up in 2014 and intensive almond orchards began sprouting in Spain's irrigated zones have pollination service arrangements for almonds become a destination for beekeepers. Much as with intensive irrigation and horticulture, Spain is the pinnacle of intensive beekeeping in many respects, just not when it comes to almonds. At least not yet.

Botanically speaking, almonds are among the small subset of crops that depend on insect pollination to bear fruit (though this is a bit more complex, as I will detail). How is it then that Spanish rainfed growers have managed to do without maintaining or importing honeybee hives? There are several explanations for this. Geography and scale play a major role. Almonds are commonly planted on land considered the most marginal, both agronomically of limited productivity and quite literally at the outskirts and edges of farm fields. Often shrublands or forests abut almond plantings, providing much better habitat for native pollinators than unbroken agricultural terrain. Like many farmers I asked about pollination, an almond grower in the foothills of Aragón emphasized the proximity to wildlands: "There are bees and wild insects here, as there's plenty of undeveloped hillside." Another grower in Andalusia said that bringing in beehives from elsewhere wasn't necessary because "the environment hasn't been harmed so much yet as to need that." By virtue of their marginality, almonds tend to be located in places where wild pollinators persist.

The fact that almond trees occupy the edges of agricultural viability has also linked almonds with small-scale honey production spatially, as they were considered some of the only productive uses of the poorest quality lands. Almond trees were valued for yielding nuts from nearly unworkable soils and rocky hillsides while bees conjured honey from the wild rosemary and thyme of the nonarable lands just beyond. The close association produced by marginality was the genesis of turrón, a famed almond-honey desert that bears protected status for its regional cultural significance along the Mediterranean coast.[53] For many rural Spanish households at the margins, beekeeping near almonds made practical sense irrespective of any anticipated pollination benefits. Many almond growers I spoke with mentioned a relative or friend who kept bees at their farm's periphery as they had for generations. In Valencia, a rainfed grower told me that he didn't personally keep bees because his "grandfather always had hives. We had them over there in the middle of those pines," just beyond the almonds. Thirty years or so prior, "a lot of people had twenty hives, ten hives, fifteen hives." The varroa mite and its associated diseases had taken a toll, however. "Back in the day, everyone had bees," an elder grower in Mallorca echoed. Though Spanish commercial beekeeping has gradually intensified and bee health has suffered, much as in the United States, sedentary beekeeping persists in rural communities. Land that is considered good for little else, not even almonds, is often good enough for bees to yield a modest honey crop.

It is not only location and spatial proximity but also the scale of rainfed almond orchards and beekeeping operations that shape the relationships between the two. Spanish beekeepers with longstanding ties to marginal lands are much smaller in scale than the full-time beekeepers who move their hives throughout the country to boost their honey flow. Managing hives is but one of many diverse income strategies. Many small-scale beekeepers sell directly to local markets while others primarily supply their friends and family. Given that almond plantings in rainfed regions tend to be on small parcels as secondary crops, along hillsides or as borders between fields, this relationship with local stationary beekeepers fits the scale of both operations. Full-time beekeepers seeking to pasture their bees would much rather find large swaths of dense forage, like irrigated melons, stone fruits, or sunflowers, than trouble themselves with hauling small numbers of hives to more widely dispersed almonds in semiarid uplands. A rainfed farmer in Murcia described scale as the key factor in whether beekeepers would bother with almond orchards. "A large farm," he explained, "interests beekeepers, but to come here where there aren't large plantations doesn't interest them." Similarly, only the largest rainfed almond farms tend to seek out honeybees due to their high dependence on almonds for income. A manager at an almond cooperative mentioned that "the largest and most professionalized" growers might seek out hives, whereas "those with three, five, or ten hectares don't bother." An almond breeder reflected on how some agronomists had been promoting the use of honeybee hives in almond orchards for quite some time, but "many plantings are small. For a farmer to get someone to bring hives is more complicated." Moving bees is logistically challenging. A commercial beekeeper wants to move hundreds or thousands of hives at a time, not a dozen hives to a small patch for a short bloom. Similarly, farmers who have a few hectares of almonds as a complement to other crops and other forms of employment are not stressing over the need to maximize production. As a business model, pollination services make sense only at a certain scale for both beekeepers and farmers.

Given limited rainfall, some rainfed growers also feared that maximizing pollination would push the trees harder than conditions allow. An agronomist working with growers explained, "There's not much water for this almond tree, so I can put bees, and there can be a huge fruit set, but later this tree is not going to be able to handle all this fruit, so it suffers, and the next year it gives you problems. In the end it's more of a problem than anything." A farmer and almond industry leader in the high plains of Andalusia described that in addition to having a relatively high population of wild bees, "in traditional zones they didn't use beehives, in part because the limitation was not that," it was rainfall. In the 1970s a few groundbreaking breeders around the Mediterranean took this logic a step further by working to create an almond variety that would not rely on insect pollination at all. To improve the resilience of almond growers in marginal landscapes, they began pursuing the elusive trait of self-compatibility.

Breeding for Bee-Free Reliability

In 1965, a young Spanish plant breeder by the name of Antonio Felipe made a bold move by pursuing research on almonds even when professionals dedicated to intensive fruit growing found it strange, given the low status of almonds as a "crop of convenience." Very few almond growers purchased nursery-grown seedlings, so why go through the trouble of breeding a new variety? It made little sense career-wise. But almonds were a key component of many struggling households on modest patches of land, and he hoped to lighten their load. After traveling around the Mediterranean hills meeting with farmers, he told me in our interview, he determined that the chief problem of almonds was their unreliability. Yields vary widely from year to year, and in addition to fluctuations in annual rainfall, the delicate bloom period is often to blame. Knowing that almond-growing households had little resources for adding fertilizer or other inputs, Felipe set out to make almond trees as resilient to external factors as possible. Their fate should not be so tightly tied to the weather or the flight of a bee.

Blooming in winter, as almonds do, is risky business even in temperate Mediterranean climes. Freezing temperatures, rains, and strong winds are fairly common during bloom times in the regions where almonds have historically been grown. The slightest shift in conditions can dramatically change the tree's trajectory for the rest of the year. Each blossom is open only for a few days, and the bloom period lasts only a few weeks. Because almonds almost universally require cross-fertilization of distinct varieties that bloom at slightly different times, there needs to be significant overlap when flowers of both trees are open. And if the weather is too cool or rainy during these precious few days, most bees don't fly. If it freezes, the blooms die all together. Some years a cold snap hits and the harvest is practically nil. Other years the sun shines and a bounty follows. Dramatic swings in productivity—a product of highly variable rainfall, freeze events, and pollination—have made almonds excellent for episodic cash infusions and practically ludicrous as a source of primary income among rural Spanish households. As a breeder, you can't change the rain or the cold, but you might be able to change how the trees fertilize their fruits.

Knowing the crucial importance of the bloom period, Felipe and those who later followed his lead pursued two main strategies to secure a more reliable crop: late flowering and self-compatibility. The later the tree bloomed, the less likely rain, frost, or cool temperatures could impair pollination and fruit set. Self-compatibility meant that successful pollination no longer depended on warm weather enticing bees out of their hives or the coordinated overlap between bloom times of distinct varieties. Combine these traits, and you can have an orchard that is planted as a single variety, blooming when weather is more favorable (as late as early April, compared to varieties blooming as early as the end of January), and no longer dependent on bee activity.

Decades of assembling collections, making crosses, scrutinizing seedlings, and then grafting desirable scions eventually achieved this ambitious goal. Almonds were long thought to be universally self-sterile, which made evolutionary sense given their origins in harsh arid environments where genetic diversity is crucial for adaptation.[54] Researchers as early as the 1930s had discovered a rare self-compatible almond variety, but because the report was published in Portuguese, it remained little known. In 1974, Antonio Felipe in collaboration with the French researcher Charles Grasselly identified a self-compatible variety from Italy called Tuono and sparked a small but dedicated network of Mediterranean plant breeders. Inspired by this work, a Spanish student by the name of Rafael Socias obtained a scholarship to study self-compatibility in almonds in a place where the crop had much greater commercial importance and public research support: the University of California Davis. After his doctoral research under the renowned American almond researcher Dale Kester, he returned to Spain and together with Felipe selected the first commercial self-compatible varieties.[55] The Guara variety, released in 1986, remains the most popular named variety throughout Spain to this day.[56] Nearly all other cultivars released since, each with its own distinctive qualities, feature the signature self-compatible and late-flowering characteristics.

Knowledge practices that prioritized resilience to external conditions, rather than productivity at all costs, gave Spanish growers a key tool for managing orchards independently from pollinators. Though it is difficult to measure, this has also likely cushioned the blow of pollinator decline for Spain's rainfed almond growers. Some Spanish agronomists I spoke with still recommended that almond farmers seek out honeybees to get the best yields because their movement of pollen between blossoms provides a boost. Few rainfed growers take the trouble.

Spanish breeders in the 1980s never anticipated that self-compatible and late-flowering varieties would someday also provide a boon to a new slate of intensive almond growers in Spain's irrigated zones. For these growers, everything was new, and the question of whether or not importing pollinators would provide a yield boost to self-compatible varieties was a subject of active debate. Some farms, particularly large plantings with farm managers reporting to investors higher up the chain, felt safest following the California model as closely as possible. Given abundant capital, and little personal experience, purchasing pollination just in case it might boost yields seemed the best bet. The managers of a corporate-owned irrigated megaplanting in Catalonia assured me that their almonds would receive ample pollination as their investors would not want to risk any limits to production. They chose to use disposable boxes of bumblebees imported from the Netherlands rather than contract with beekeepers, because they were familiar with this practice in commercial greenhouses in the region. A nearby farmer with a more modestly scaled orchard that he owned and operated, said he actively seeks out beekeepers to ensure pollination but has never had to pay. Another irrigated grower in the area said bees didn't seem necessary and he wanted to see more scientific proof. A farmer near Zaragoza with both intensive and "superintensive" plantings (growing trees in tightly packed hedgerows) was distrustful of recom-mendations to bring in bees to boost the production of self-compatible variet-ies. He said he had been promised a 5–10 percent yield increase, so he paid for bumblebee hives the year prior but saw no results. A website aiming to matchmake between farmers and beekeepers launched in 2022 but seems to have gained little traction.[57] A history of breeding for marginal environments and risk reduction had serendipitously provided a form of resilience to even those intensive growers pushing for maximum productivity.

The Pursuit of Independence

Given the well-known popularity of self-compatible varieties in Spain since the 1980s, why is it that almonds are still widely categorized as "requiring" insect pollination? If self-compatibility is a clear option, why do California growers still "need" to truck nearly all the honeybees in the continental United States to their orchards every year at significant economic and biophysical cost? California researchers had been interested in self-compatibility since Kester's work in the 1960s, but it was never a top priority for breeders. Not until the price of rent-ing honeybees started to sting for growers. Yet despite ravenous demand for a new suite of self-compatible varieties, the dream of bee-free almond growing in California has never come to fruition. The economics of intensive production, more than any biological constraint, has created a dynamic in which no amount of breeding appears able to disentangle the almond and the bee.

Around 2006, the cost of bringing beehives to pollinate California's almonds jumped from around $150 to $350 per acre as beekeepers resorted to more inten-sive tactics for stabilizing hive health.[58] It was perfect timing for the release of the Independence almond, a variety developed over decades by the private breed-ing company Zaiger Genetics, whose signature trait is self-compatibility. As the prices of almonds hit record highs, sparking a wave of new plantings in the early

2010s, the Independence accounted for roughly a quarter of nursery sales by 2016, second only to the standard-bearer Nonpareil.[59] Such rapid adoption was a raving success in a capital-intensive industry where growers are cautious about an investment they will have to live with for twenty or more years. The Independence hadn't even been around long enough to provide data on its full life cycle, but growers were willing to take a risk in hopes of shedding their ballooning bill for pollination services. Planting blocks of a single variety also offered to reduce the hassle of managing alternating rows of different varieties with distinct needs and management calendars. The age of the self-compatible Independence almond had arrived.

The promised independence, however, was never achieved. Several key factors thwarted the potential for California almonds to go honeybee-free. The first was the scale of almond growing operations. Solid blocks of Independence, I was told, were simply unfeasible for large operations with many thousands of acres. The variable timings of sprays and harvest activities required by two or more varieties allowed large growers to make better use of staff and equipment over the course of the season. "You can't harvest them all, that's thousands of acres at one time," a Stanislaus County grower explained. "The trees all mature and the nuts have to be harvested at one time, and that's an advantage of Independence on a twenty-acre lot. You don't have to go and bring the equipment in there twice," for two cross-pollinating varieties. Independence was better suited to small growers looking to simplify and hedge against the risks of uncertain pollination futures. Even more limiting for the adoption of Independence was the market. Buyers still wanted the appearance and taste of the Nonpareil nuts they had become accustomed to (see chapter 1), and Independence couldn't match its price. Even with all the enthusiasm for self-compatible varieties, in 2022 they only accounted for 9.8 percent of total acreage, and the percentage of new plantings seems to have plateaued around 25 percent. It served a niche but was unlikely to dominate the industry.

Market pressures also meant that even Independence "needed" bees. With almonds as profitable as they were, it still paid for growers to place hives in the orchard to boost yields by a few percentage points. Farm advisers lacking robust scientific data recommended one hive per acre rather than two, just to be on the safe side, as rumors floated that plantings without hives had underperformed. Some in the industry saw this not only as rooted in scientific rationale but also as a way of relieving controversy. Farmers with long-established varieties felt that neighboring plantings of Independence were now freeloading off their expensive imported hives and hurting their harvest. After all, bees do not respect property lines, and they were likely to fly over to the neighboring Independence, reducing time spent among the trees of their paying customer. In 2020, researchers led by a Chilean-based group confirmed the commonly held view that Independence benefited from bees, reporting 20 percent yield increases.[60] Sacrificing such potential would be economically unacceptable.

Even if the industry moves toward a plurality of self-compatible varieties with variable harvest dates, millions of managed honeybee hives will still be part of the picture. The Almond Board and its plant breeder partners anticipate that new self-compatible varieties will more likely serve as pollenizers interplanted with the reigning champion Nonpareil. Breeding for self-compatibility in California might bring the costs of hive rentals down a notch, but the mass pollination migration is likely to remain deeply entrenched. Almond orchards reached a point in 2021 where even 100 percent of the honeybees in the United States would not meet their demand for two hives an acre.[61] Self-compatible varieties are effectively enabling almond orchards to continue expanding while beekeepers scramble to maintain the same number of hives healthy enough for the pollination job.

DENATURALIZING POLLINATOR DEPENDENCE

The largest managed migration on earth is far from a matter of biological necessity, or evolutionary destiny; it is the confluence of two colliding technological treadmills of capitalist agriculture. Yet agronomic reports and media coverage perpetuate the idea that almonds are deeply and irrevocably pollinator dependent. These are economic relations masquerading as biological facts. Looking at California and Spain side by side, we see how intensification is the real root of the almond-bee industrial symbiosis. By contrast, the material and social conditions of marginality foster resilience.

In both California and Spain, beekeeping became untethered from fixed locations with the availability of fossil-fueled transportation. Beekeeping to produce commercial honey became a specialized endeavor, rather than a complementary activity. As the scale of operations increased, beekeepers sought out large sites with abundant, often irrigated, flowering crops. In California this meant monoculture almonds, cherries, and later apples or canola, but in Spain it meant melons or sunflowers. Migration was a form of intensification allowing beekeepers to keep pace with the mounting pressures of disease and chemical agriculture. California growers heeded the advice that agronomists promoted to import honeybees during the bloom and gradually began paying to bring hives from farther afield. Beekeepers in turn became increasingly dependent on almond pollination income as their hives endured blow after blow from pesticides, poor nutrition, parasites, and pathogens. Prioritizing pollination services, rather than honey, has radically reoriented the work of beekeepers and intensified dependence on yet another slew of industrial agricultural inputs—corn syrup, soy protein, and miticides—each with their own ecological ramifications. Though hive numbers remain high as the money is good, honeybee health continues to suffer. The almond pollination event likely does more harm than good for the bees, but their destinies are now intertwined. If the almond industry takes a dive, the beekeeping industry will too.

Meanwhile the vast rainfed areas planted with almonds throughout Spain have remained largely shielded from such codependency. By occupying marginal spaces, these trees benefit from proximity to habitats conducive to wild pollinators. Often this proximity also has made beekeeping by almond-growing households or their neighbors a practical pairing. Rainfed growers see the importation of honeybee hives as indicative of a more industrial scale and style of farming. The small-scale, nonintensive, and often irregular contexts of almond orchards make coordinating a visit from a migratory beekeeper more hassle than it's worth. When a local beekeeper seeks out a place for their hives and finds a spot on almond-growing land, all the better, but the low-stakes exchange is expected to net a few honey jars, not make or break an almond harvest. Breeders with the interests of struggling rainfed growers at heart eventually provided a genetic solution, self-compatible almonds, which for many offered resilience in the face of bad weather and limited pollinator activity.

The subsequent development of self-compatible almonds in California demonstrates the fallibility of biological explanations of pollinator dependence. Even when almond trees do not physically require an insect to transport pollen, dependence on honeybees merely shifts by degree. To decline pollination is to leave money on the table. The dependency is not in the plant's physiology, it is in the market.

What else might we learn by challenging the presumed dependence on imported pollinators? Instead of asking whether or not a plant requires imported honeybees, we should ask how and why this dependence has come about. Who is truly benefiting, and what are the long-term ecological consequences? Feeding honeybees corn syrup and soy, while dousing them in ever-higher doses of mite-killing chemicals, puts honeybees and the lives with which they are entangled in a precarious position. This is not about "saving" the bees. In North America, honeybees are livestock without a niche in native ecosystems. Somewhat akin to cattle, they have not evolved here, but they can be managed holistically, or they can be packed into a feedlot. Addressing the pollinator crisis, for both native and managed species, takes coordinated efforts to provide more diverse flowering landscapes, reduce reliance on pesticides, and probably truck around a lot less bees. It may also mean fewer, healthier bees. What is apparent is that we cannot protect pollinator health without changing the market conditions of agriculture.

A pattern is clear. Almond growers, once largely indifferent to beekeeping in their orchards, have been pressured by both their scientific advisers and market pressures to become deeply dependent on pollination services. Now beekeepers, who once enjoyed a lifestyle where their hives were simply left to their own devices for much of the year, are under pressure to feed their bees to meet grower expectations. Bees "need" corn syrup about as much as almond trees "need" imported honeybees. The dependence has been produced by a feedback loop between a landscape no longer hospitable to insect bodies and a narrow agronomic focus on productivity rather than persistence. The rise of pollination services around the world tells us little about the inherent relations between flowering plants and their insect partners and much more about the dire state of our agroecosystems. To get at the root of these challenges requires reimagining claims to biological dependence less as fixed conditions and more as openings for intervention.

4

———

Space

Creeping Toward Precarity

How do we make judgments about where almonds (or any other crop for that matter) should grow? California's almond boom drew public outcry not only because of when it was happening, during a major drought, but because of where: in the so-called desert. What most casual observers did not realize is that the bulk of California's almond production not only occurs in exceptionally arid climes, but it has also gradually shifted into drier regions over time. This is no accident. Growing almonds in the desert makes good agronomic sense. In the pursuit of efficiency—maximizing yields per unit of land, water, or nitrogen—arid lands provide quasi-factory levels of control. The steady drip of microirrigation (which distributes a custom cocktail of fertilizers and pesticides known as fertigation and chemigation) is much easier to fine-tune without sporadic and unpredictable storms getting in the way. Frost and humidity pose risks for developing fruits, and spring rains can limit pollinator activity. The expansion of an intensively irrigated crop in lands nearly devoid of rain is not only a matter of water being cheap and under-regulated. It is also a direct result of the industrial ideal in agriculture: maximizing uniformity and control.[1]

While the boom of the 2010s merely accelerated the long-term shift in California's production from north to south (and thus wetter to drier conditions) along the Central Valley, the flush of investment radically and rapidly reoriented Spain's almond geography. After centuries at the margins of agricultural production as a complimentary crop, almonds have now descended into intensively irrigated valleys and plains. These regions are already suffering from overexploitation and contamination of groundwater, problems likely to become more intractable with the transition to a high-value permanent crop. Much like the shift to increasingly

97

arid production in California, Spain's reshuffled almond geography is entirely rational according to agronomic logic. It is heralded as a move toward productive efficiency, offering high yields per hectare and high profitability. Many agronomists and industry members I met appeared to be breathing a sigh of relief that finally almond orchards were being planted in the "right" places, where they could reach their full potential. If your objective is to optimize a single crop's profitability, the benefits are clear. But does this mean then that all crops should ideally be grown where they can be pushed to their physiological limits to maximize production? Especially when we know that overproduction is responsible for so many of industrial agriculture's social and ecological failures, not to mention problems for long-term profitability? What happens to the landscapes where almonds have long served as the only viable option, or a last resort for maintaining relations with the land?

This chapter argues that agronomic ways of knowing are inherently spatial. While seemingly limited to the farm field, they have deep implications for how agriculture relates to broader socioecological geographies. By ignoring these geographies, the agronomic fixation on productive efficiency actively produces precarity. These ways of knowing are not merely scientific, they are political economic, fueled by an influx of capital and structured to benefit those in a position to profit. There is little room in such accounting for the losses incurred in the process. Yet those who break from this narrow approach to valuing agriculture and find the arboreal ingenuity of almond trees may very well anchor a reimagination of the rural.

CALIFORNIA: THE ATTRACTION TO ARIDITY

California's Central Valley, a strip of farmland that is 40–60 miles wide (60–100 kilometers) and increasingly carpeted by the deep green of irrigated almond orchard canopies, stretches 450 miles (720 kilometers) from north to south. The valley's climate is a gradual gradient from cooler, wetter weather in the north (15–30 inches, or 381–762 millimeters, per year) toward hotter, drier conditions in the south (5–15 inches, or 127–381 millimeters, per year). Looking at the history of the state's shifting geography of almond production is like watching a slow march southward along this narrow strip of earth. In 1950, roughly half of almond production came from the Sacramento Valley to the north.[2] In 2022, these same northern counties yielded just 13 percent of the total output.[3] The irrigation made available by the California Water Project enabled a surge in almond plantings in a "more favorable climate," which is to say warmer and drier.[4] This shift has enormous implications for the valley's ecology and the state's water management. Almonds under such conditions require more water not only due to limited rainfall but also because plants transpire more quickly as evaporation pulls moisture skyward. Yet the region has little water to give. The town of Bakersfield, in the

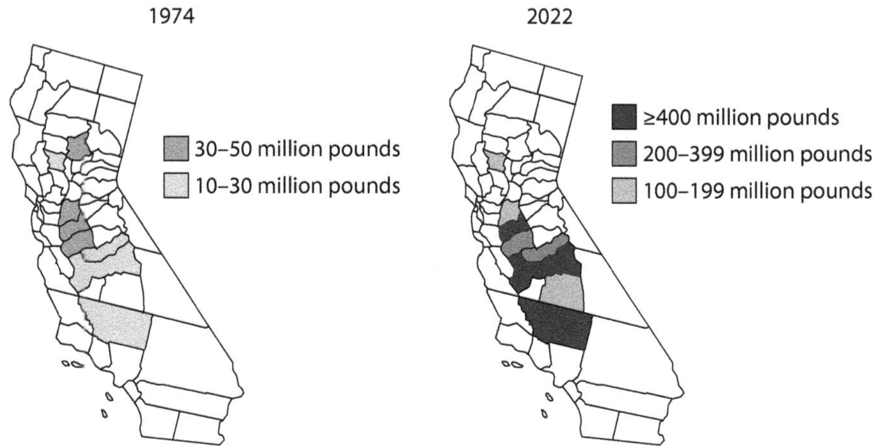

MAP 1. Top almond-producing counties in California. Source data: USDA (1974), Almond Board of California (2022). Map by author.

heart of today's almond country, has averaged under six inches of rainfall per year for the last decade. The almond industry describes this southern portion of the valley as more efficient, not only in terms of yield per acre but also in terms of water use or "crop per drop." Growing in a near desert is, in agronomic efficiency terms, an environmental triumph.

Aridity offers the closest thing to a blank slate upon which the industrial agricultural ideal can unfold. Compared to the precision application of water on an intensive almond plantation, rain is highly inefficient. Microirrigation techniques like drip lines or microsprinklers are designed to release only as much water as is optimal for plant growth at precisely timed intervals to maximize absorption. On California almond farms, these systems typically account for changes in weather on a daily or hourly basis, thanks to an array of state-supported weather stations and publicly funded crop science, ensuring that as temperature and humidity changes, the optimal irrigation level adjusts accordingly. Irrigation systems are also typically zoned to account for variation across large acreage such that the irrigation patterns for each block of trees, or perhaps someday each individual tree, can be customized (though these variable rate systems have their limits). Unlike the blanket distribution of rainfall which needlessly dampens the windrows between trees and promotes weed growth, precision irrigation directs water only to the root zone. And while nutrient or pesticide applications do not factor explicitly into the "crop per drop" ethos, it is a marvel of modern engineering that this entire system not only irrigates but fertilizes the crop and suppresses its soil pathogens simultaneously with little additional effort.

Rain, by contrast, rarely conforms to the optimization of an individual species, let alone one that did not evolve in the climatic conditions where it falls.

Rain nourishes a region's entire ecology, in which noncultivated species depend on variability across multiple scales.[5] For a farmer, natural rainfall often provides too much water or too little. It might flood fields, clogging up the wheels of machinery, or create fast-moving rivulets that add chaotic contours to laser-leveled topographies. Rain can wash away fertilizer before it has been absorbed by the tree, demanding either additional costs for more materials or additional risk of suboptimal yields. The lost nutrients then flow onward to contaminate streams or trickle down into groundwater. When rains come too early or too late or too sporadically, harvests are likely to suffer. Rain is known to withdraw into the skies for weeks or more, or to reluctantly release a meager supply. Prayers not just for rain but for the right kind of rain are surely one of the oldest and most universal agricultural traditions in the world.

Critically for almonds and other insect-pollinated crops, rain during the bloom can deter honeybees from making their rounds. Hives imported from across the country making up nearly a third of growers' production costs may sit idle through the storm. If a cold snap combines with humid air, even for just a few hours, frost forming inside the delicate blossoms can wipe out a substantial portion of the crop. Unfortunately for California's almond growers, the sensitive bloom time falls in late January through February, when rains are quite common. To protect the bloom and ensure good pollination, land in a more arid climate is a much safer bet.

Rain also brings with it humidity, and with humidity comes all manner of pests. Insects often breed and feed in moist environments, nibbling at leaves or burrowing into fruits. Fungal spores readily proliferate in moist environments, colonizing budding fruits, coating leaves or rotting tree trunks. Even without rain, the dark leafy density of heavily fertilized almond plantations transpiring each day risks fungal disease. What if one could enjoy all the benefits of water directly fed to a tree's roots without any of the humidity or irregularities of rain falling from the sky? What if one could irrigate and fertilize in precise amounts without worrying about the vagaries of the climate? The factory ideal of industrial farming persistently struggles against the biophysical dynamism of the field,[6] but arid lands are about as close to a factory's sterility and exactitude as an outdoor farm can get.

The trend is clear to those who look at the industry as a whole. One of the many California almond growers I spoke with, who also earned income as a real estate consultant, explained why moving production southward makes economic sense. Sitting in Bakersfield, a city at the heart of the top almond-producing county, he explained:

> We're farming in a desert, and that's a really good thing agronomically in the sense that you don't have the pressure of disease issues that are high humidity, high rainfall, high whatever . . . one of the reasons that the almond yields have gone from what they used to be thirty-five years ago. When I first started the business, that particular business to what there today is, you've really seen a migration of that production within California . . . they found that when the almonds migrated north and south,

from there during the '60s, '70s, and '80s, that the yields even got better. Why did they get better? They got better because bees have a very strong union, and bees don't fly when the wind is blowing or it's foggy or cold or rainy. . . . That's where the economics of that seem to be the best. Which is south of Modesto. The further south, typically the better that gets going back to the rain patterns in California, the rain and wind.

Not only have yields increased and costs gone down due to the dry climate, but production statewide has also become much more consistent. "If you looked at the percentage fluctuation crops, it was mainly weather driven. . . . The south is essentially being more reliable than the north because of weather." That reliability is a boon to growers as well as processors and buyers across the value chain.

Rain, long the lifeblood of all agricultural societies, can become quite a nuisance under industrial conditions. Of course, agriculture that avoids rain is only conceivable when critical farm inputs can be reliably imported. Water is drawn from distant rivers or pumped from increasing depths. Honeybees are shipped on semitrucks from more humid climes where fields of wildflowers or other crops host them for the other forty-nine weeks of the year. California's almond farmers and researchers approach arid land much like the growth medium of a hydroponic greenhouse, a set of relatively inert physical containers that support the transformation of inputs into outputs. Given the demands for precision and intensity on an industrial farm, dry air brings peace of mind. Warmth brings additional advantages. For almonds in California, the San Joaquin Valley offers a longer growing season. Trees have more time to grow and develop before their annual dormancy. Higher temperatures earlier in the year also reduce the risk of frost during bloom, one of the frustratingly uncontrollable aspects of production. In the desert, efficiency thrives.

The political economic history of land plays a role in the rise of California's intensive arid agriculture too. The narrative of growing almonds in a "desert" has a touch of irony for those familiar with the southern San Joaquin Valley's ecology. What today appears as a parched landscape was once a vibrant wetland, home to Tulare Lake, the largest freshwater lake west of the Mississippi and the heart of Indigenous life prior to colonial conquest.[7] The wetland was intentionally drained and the lake's tributaries diverted for agriculture, acts that were as economically as racially motivated, given prominent claims that wetlands harbored inferior peoples and threatened white bodies.[8] Such assertions directly undergirded policies to "reclaim" wetlands, including the Swamp Lands Act of 1850, which allowed states to sell off large land parcels if new owners could guarantee drainage and productive agricultural use.

The swampy history of the arid San Joaquin Valley underscores the role of the state in the success of the region's almond farms with respect to their northern counterparts. Economies of scale favor the large farms of the southern counties, which can reach hundreds of thousands of acres whereas northern orchards often hover around one hundred acres or less. Public land sales, under the Swamp Lands Act, the Desert Lands Act, and other mechanisms, were a boon to the tycoons of the late nineteenth century, enabling vast landholdings that remain imprinted on the landscape today. Whereas the Sacramento Valley to the north had a population of settlers who established household-scale homesteads provisioning the mining boom, the San Joaquin Valley to the south was the investment playground of wealthy urban capitalists who even used their political influence to forcibly remove would-be homesteaders.[9] Long before irrigation infrastructures or almonds came to the San Joaquin Valley, the state and its alliance with capitalist interests enabled a concentration of wealth and territory, which contributed to economic advantage. Thus the agronomic benefits of aridity and warmth, enabled by state-backed irrigation infrastructure and land appropriation, have made the San Joaquin Valley an almond-producing machine.

At no time in the California almond industry's history was water efficiency a driving factor in this southward shift; however, the logic of efficiency became a strong defensive strategy in the face of public scrutiny. Almond production in California increased its efficiency in terms of "crop per drop" by over 33 percent in twenty years, the Almond Board proudly reported during the height of the 2014 drought. Year after year, on average, growers are producing more almonds from each unit of water. What the industry group did not report is that, during this same period, growers had increased their water applications per acre by 37.4 percent. The orchards had become more efficient, but overall water use had not declined. Exactly the opposite. Growers were simply squeezing more almonds out of every acre. While somewhat counterintuitive, this "crop per drop" narrative perfectly reflects the preference for arid lands. Precision irrigation and nitrogen mean that every drop goes toward plant productivity. Natural rainfall interferes with such a strictly on-demand system. And when it comes to resource efficiency, rain is an inconvenience, failing to conform to precision-oriented industrial logics.

Like a Moth to a Flame

Economic and resource efficiency are abstractions, metrics imagined to exist outside of any specific place. According to neoliberal logic, the invisible hand of the market guides agricultural production to its most profitable locales.[10] In the case of California almond production, short-term agronomic optimization layered onto a history of settler-state land dealings and water infrastructures has resulted in a spatial shift toward increasing precarity.[11] The very counties where almonds are booming are those suffering the worst impacts of aquifer overdraft, contamination, and climate threat. The almond industry is by no means the primary cause of these impacts, but it has worsened them by being drawn to aridity, like a moth to a flame. By excluding any awareness of place-based concerns, efficiency logics can actually encourage and justify spatial shifts toward problematic places.

California almond growers are feeling the strain of groundwater dependence as dry years limit surface water supplies and water tables fall. Wells that used to pump continuously have become unreliable or run dry completely. I spoke with a grower who now manages several farms for institutional investors and described the mounting maintenance headaches. "Wells going dry. That was the biggest thing, but there was . . . always a well problem. They weren't always going dry, but they weren't producing what they should . . . needing service, perforations plugging." During the most recent drought the costs to drill a deep well could surpass a million dollars, and drilling companies were booking more than a year in advance. Materials, motors, and other pump equipment were in short supply. Deeper wells complicate matters further, he explained. "As pumping increased, water levels dropped, water quality dropped, more and more salts were being pulled out of the ground, and we weren't able to irrigate or have enough wind or rain to leach those or dilute them. As a result, the salts were building up in the soils, and it takes years of quality water to flush that out. We're going to have residual salts in our soil for several years going forward. . . . We think there'll be another drought and this cycle will repeat itself." While each well site has distinctive characteristics, deepened wells in this region often show higher salt concentrations resulting from legacies of fertilization.

Pumping from greater depths requires an immense amount of energy. Water is heavy. A study commissioned by the California Public Utilities Commission reported that irrigated farms increased energy use by about 25 percent on average during drought years.[12] This means that groundwater reliance and depletion ratchet up fossil fuel use as well, as roughly half the state's energy during the 2012–16 drought came from natural gas. Farmers meanwhile have bemoaned the rise in energy prices compounding this economic burden. A large-scale grower in Kern County who began planting almonds in the 1970s explained his surprise at the mounting costs of energy. "We used to pump back when almond production first started. You were talking about one cent per kilowatt hour to pump water. Now it's in the twenties."

Drought is an expensive problem for growers. New or deepened wells, high prices on water markets, and mounting energy bills all eat into profits. Strangely, abundant water immediately following dry years can actually hurt growers too. The profitability of almonds has thus far made up for the added costs of irrigating during drought years, and acreage has continued to increase. A farm manager at a large operation near Fresno described the fears that more water would mean the boom in production would be even greater, putting a damper on prices. "This season, I think there's like a 1.1 million acres yielding. Next season, they're talking like 1.3 million. So that's a scary number if we have good weather. And we had a decent water allocation this year, so it's shaping up to be a pretty massive crop next year. It's definitely outpacing demand in the short term." The flush of readily available water paradoxically threatened to shrink profits that year. "I think in the long term, people are pretty bullish though," he added. The bets being made were on the order of decades, not years.

When California passed its landmark Sustainable Groundwater Management Act (SGMA) in 2014, many farmers recognized just how fragile their deepening dependence on groundwater had become. The placeless productivity paradigm had trapped them. The plan calls for bringing aquifers into balance by 2040. Balance will be quite the undertaking in prime almond-growing areas like Kern County, where rainfall averages hover around six inches and irrigated orchards currently draw around fifty-six inches, nearly ten times as much. For the first time in a century, the broader hydroecological context of the farm is beginning to matter. As one senior farm manager for a large farming corporation put it, "there's going to be severe pain. You have a resource that's freshest to mankind. . . . They're not going to let a few abuse it. They won't. That's what we've been doing." Many farms are buying land for the associated water rights, either to irrigate their crops or in anticipation of high water prices. Farmers are also actively shaping the water management plans, with some critics pointing out that the law's emphasis on local governance has effectively allowed the foxes to guard the henhouse.[13] Another longtime Kern County grower and real estate professional explained, "If we don't come together and if we don't fight for our water, then we will see this acreage be cut back quite a bit. If the SGMA goes as planned and they say, we're going to get two acre-feet, well, we'll be growing half the acres that we can grow. . . . I think, unfortunately we need to get involved with those type of organizations to go out and come together to have a voice that we can be heard." If other groundwater-dependent agricultural economies are any indication, the battles over implementation of SGMA will be fierce and local leaders may lack any real incentive to conduct strong enforcement.[14] In fact, in April of 2024, California's Water Resources Control Board rejected locally developed plans from one prime almond-growing region as insufficient and shifted control over to the state. Time will tell if California can demonstrate a model for effective enforcement of groundwater law.[15]

Water is not the only source of precarity associated with the southward shift of California's almond production. Pest pressures, and the toxic materials deployed to suppress them, have also increased. A robust analysis of pesticide application trends demonstrates a clear gradient in increased use of insecticides and herbicides across the valley, with the southernmost orchards using more than triple the insecticide application per acre of their northern counterparts.[16] There are several likely contributing factors to this. A warmer climate with an extended growing season for the almond trees facilitates a longer reproductive season for insect pests as well. Primary insect pest populations for almonds, such as navel orangeworm and leaf-footed bugs, tend to show population spikes during a number of discreet emergence events, or "flights," per year. Pesticide sprays are often planned according to "degree days," a calculation combining temperature measurements and the pests' life cycle pattern. While this method has been reliable for northern growers, several interviewees in the southern region described so many flights that the pressure is effectively continuous. One Kern County grower who also manages investments for institutional clients explained, "You're watching your degree days. You're hitting your spray timings. You spray exactly when the optimum time is, and yet you have a failure. That's just because the pressure's so high. You don't have those definite points in the life cycle anymore. You have multiple generations overlapping each other, so it's a constant wave of pressure versus peaks and valleys."

Not only are insects able to up their number of reproductive cycles, they also thrive on more densely packed and continuous orchards in these southern stretches. Following University of California recommendations, growers have largely abandoned annual pruning beyond the tree's initial establishment, sacrificing a modest amount of productivity while saving a significant amount on labor costs.[17] At the same time, they have been encouraged to plant more densely. The result is a thick, virtually uninterrupted tree canopy providing more biomass which can house and feed larger insect populations. Such densities pose challenges in wetter northern counties, given a higher propensity toward fungal disease, but the dry desert-like air makes fungal growth less threatening. The popularity of almond production in the arid south, given its high productivity and profitability, has also made the landscape more homogeneous. Whereas the northern valley hosts smaller farm parcels and maintains a relatively diverse array of nut orchards along with stone fruits, vines, and row crops, in the south it is now common to find almonds as far as the eye can see in all directions. Thus the southward march of the almond industry over recent decades has nurtured pest populations and ratcheted up demand for the poisons to suppress them. Such chemical dependency is harmful to human lives and broader ecological processes. It is also highly precarious. The active agents in these products are becoming less effective over time due to pest resistance,[18] and their toxicities are gradually being recognized and restricted by regulators. Warmer and more arid climes have allowed southern almond orchards to achieve unprecedented productivity, but they do so on borrowed time and at substantial cost.

Precarity for Whom?

The geographic shift of the California almond industry toward the arid southern San Joaquin Valley makes agronomic and economic sense, but only if the logic is contained within the narrow bounds of the farm. Groundwater legislation has begun to bring the hydrologic context in which a farm sits more directly into the calculus. This may bring financial precarity for some farms, particularly smaller operations likely to be outbid on water or land. But no one I interviewed seemed terribly worried about their personal livelihood being in significant jeopardy. Among those I spoke with, growers on smaller farms (described by those in the industry as under three hundred acres) all had a stable source of nonagricultural income in their households, such as working at a local bank or having a private medical practice. Those with midsize operations committed to agricultural careers had typically been diversifying their incomes into real estate or consulting. Corporate-styled megafarms, regardless of if they were owned by a single family or an investor group, hedged their bets across an array of crops, properties, and revenue streams, often working toward vertical integration to capture more value. Overall, farmers I spoke with expected that agriculture's footprint in the Central Valley would likely shrink and change in character and that the farming opportunities for the next generation would be more limited, but no one feared that their family would sink into financial ruin. At 2023 prices, just fifty-four acres of irrigated California agricultural land (on average) would be worth more than a million dollars.[19] Anyone who owned California property knew they would land on their feet.

Perhaps unsurprisingly, most of the precarity resulting from the geographic shift toward aridity has been shouldered by those outside the farm gates. The region's most marginalized population, Latine farmworkers and their communities, are most at risk of losing the water supply to their homes. In 2016, the unincorporated town of East Porterville presented a graphic illustration of such inequities and attracted a surge of media attention. As growers scrambled to deepen their irrigation wells, more than 500 domestic wells ran dry in an area with approximately 1,800 households.[20] Houses in East Porterville did not have access to the municipal supply in Porterville proper, making the community entirely reliant on these sources. Like other low-income, predominantly Latine, unincorporated communities, this historic process of exclusion from municipal services was an unmistakable reflection of the racially segregated geography of the valley.[21] The lifelong lived experience of cultural and political marginalization could not easily be remedied by technical means. Years later, when subsidies for connection to municipal supplies eventually became available, many East Porterville residents chose not to engage with state agencies because they feared deportation or had heard rumors that they might risk losing custody of their children.[22] The suffering caused by failing wells went beyond the struggles of acquiring water, revealing how the day-to-day challenges of cooking and bathing with scant supplies intersected with a political landscape of neglect and intimidation.

The water injustice evident in East Porterville laid bare a pattern throughout the region. Four years later, a 2020 hydrologic study recorded nearly one in five wells running dry during drought events throughout the Central Valley.[23] A disproportionate number of these incidents cut off water to homes, which have more shallow and vulnerable wells.[24] Overpumping not only jeopardized the water supply of those dependent on wells, it also intensified the agrichemical contamination of groundwater for those served by municipal sources. Nitrate contamination from heavy fertilizer applications and livestock is a pervasive problem linked with methemoglobinemia ("blue baby syndrome"), reproductive harm, thyroid abnormalities, and cancer.[25] Water systems serving areas with predominantly Latine households, low rates of home ownership, and smaller communities are disproportionately exposed to these hazards.[26] This burden of toxicity is made much worse when pumping ramps up during droughts, with high-nitrate groundwater being detected at four to five times the rate of nondrought periods.[27] Naturally occurring contaminants like arsenic and manganese become more problematic due to overpumping as well, especially for disadvantaged populations.[28] While sources and geologic factors are complex, contamination problems that are relatively minor in the northern Central Valley generally become more acute as one moves toward the southern counties where aquifers are more strained and agrichemical use has intensified.

Toxic exposures are of course not limited to groundwater. The pesticide intensity of highly productive orchards in southern counties also heightens health risks for workers and surrounding communities. Warm weather and dense, continuous orchard habitat have boosted pest populations, deepening dependence on agrichemicals. For decades almond growers relied heavily on the insecticide chlorpyrifos, a neurotoxin known to cause brain damage in young children.[29] It was banned in California in 2020 and by 2023 was banned nationally on food and feed crops, save for certain exceptions.[30] A study of application rates in 2003–2007 shows that the southernmost counties applied the toxin at triple the rate of their northern counterparts.[31] The Almond Board of California, in anticipation of regulation, has been proactive in encouraging growers to shift away from the most acutely harmful organophosphate insecticides, like chlorpyrifos, replacing them instead with pyrethroids. While far less acutely toxic to mammals (including human beings) and birds, these products continue to jeopardize many insect populations, including pollinators and others viewed as beneficial for agriculture.[32] Run-off also carries the compound into aquatic ecosystems where it can be fatal to fish and amphibians, as well as alter their behavior and reproductive success at lower doses.[33] Regardless of the transition to new products, however, the trend holds strong that the southern counties are deeply dependent on higher loads of pesticides.

The almond industry is by no means the sole or even primary contributor to the environmental injuries and injustices experienced by communities in California's Central Valley. My argument is that almond production has moved *toward* these injustices. This is more than an unfortunate coincidence. Growing in warm, dry climes, provided irrigation water extracted from elsewhere, is very profitable. The same agronomic logic that makes growing in arid environments attractive also makes it complicit in very real and unequally distributed harm. By imagining that farms are merely a set of inputs and outputs—rather than embedded in ecosystems, hydrologic systems, and social landscapes—productionist agricultural sciences have cultivated a culture of willful ignorance. The ability to evade recognition of broader relations reflects the way racial capitalism feeds on divisions. The disproportionate burdens placed on the poor and on communities of color expose their relegation to a position of disposability.[34] The geographic shift enabling more efficient production systems, no matter how much the industry might brag about pumping out more crop per drop, has drawn almond production toward locations of more pronounced socioecological harm.

SPAIN: RUSHING TOWARD THE IRRIGATED PLAINS

While the California almond industry's geographic shift took place over many decades, in Spain the boom of the last ten years has made for a much more rapid and dramatic geographic shift. Whereas until recently orchards had been almost exclusively rainfed, the profitability of irrigated plantings (alongside the decline in grain prices) has drawn almond orchards into intensively irrigated districts served by major dams and diversion infrastructures. Lush, irrigated almonds fanning out across the plains have created a landscape that would have been considered absurd just a few years before. Many long-time rainfed growers regarded these plantings with a degree of awestruck horror. While strong prices fostered hopes that almond orchards might anchor the revitalization of rainfed zones, the towering piles of almonds churned out by intensive production also cast an ominous shadow of economic precarity. Farmers know the boom-and-bust cycle all too well, and the market can be particularly unkind to those at the margins.

During my fieldwork in 2018, the notion of where almonds "should" grow in Spain was undergoing rapid transformation in the face of new political economic configurations. The mixture of eagerness and unease with which this was occurring serves as a helpful reminder that spatial understandings of agriculture are always contingent. There was nothing natural about Spain's almond production shifting to inland irrigated areas, yet at the same time many in the almond industry found justifications for why this was where almonds should have been grown all along. Almond trees, they explained, had finally taken on their rightful identity as a "frutal," or fruit tree. This reclassification meant they would no longer be a peripheral, low-input activity for rural households in rainfed zones but rather had entered the realm of large-scale commercial agribusiness. To consider almonds as fruit trees meant supplying abundant irrigation, fertilizer, and pesticides and utilizing high-volume harvesting machinery. These practices require large, flat expanses with access to plentiful water. Almond production, once considered a perfect match for rainfed highlands, unusual niches, and uneven grade, swiftly took root in the vastness of the irrigated lowlands.

Unsurprisingly, the wide valleys where intensive almonds sprouted up were some of the most burdened by decades of industrial agriculture, including over-exploited and contaminated aquifers, pesticide toxicities, and biodiversity decline. For almonds to be refashioned as an engine of rapid capital accumulation, they not only had to take on industrial practices (like pesticide dependence) that produce precarity, they also had to move *toward* spaces already plagued by precarity. In the

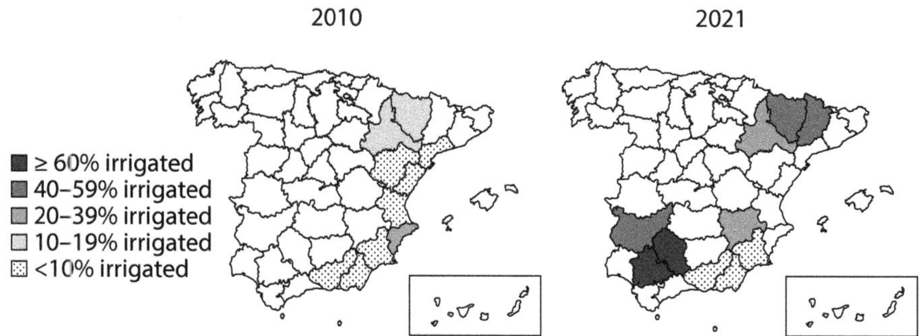

MAP 2. Top almond-producing provinces in Spain. Source data: Ministerio de Agricultura, Pesca y Alimentación. Map by author.

cases of recently built canals, almond production became a vehicle for land speculation, feeding off existing economic instability.[35] The geographic reconfiguration also risked leaving precarity in its wake. While some rainfed farmers staked their hopes on almond-centered futures, many feared that social and ecological dislocations of abandonment would eventually follow.

Boom Times Barrel Toward Precarity

Spain's almond geography has always been dynamic, though previous shifts pale in comparison with the rapid relocation of production in recent years. In the early twentieth century, almonds became popular among a class of newly landowning farmers with small plots, as almonds were low maintenance, easy to intercrop, and compatible with wage work. Government subsidies in the 1970s prompted a wave of plantings, expanding monocultures in rainfed zones and incentivizing plowing, which likely worsened erosion in many sloped areas.[36] In the 1990s, the availability of late-flowering varieties gradually extended rainfed production inland from the coasts, especially into cooler northern climes like the foothills of Aragón. Over the latter half of the twentieth century, tourism edged out agricultural land uses near coastal destinations, yet rainfed almond production largely persisted along the rugged topography hugging the Mediterranean. No prior shift, however, could compare in speed and scale with the almond boom of the 2010s.

After decades of relative stagnation, almond prices shot to unbelievable highs, rising from 2.11 euros per kilogram in 2011 to 7.6 euros in 2015 on average, a 360 percent jump.[37] Legendary reports of selling for 8 euros spread quickly among farmers, transforming a modest boom into a full-fledged planting frenzy. Such unprecedented profitability was largely the result of immense marketing resources mobilized by the Almond Board of California to drive global demand, in combination with dietary trends in the United States and Europe toward low-carb, protein rich, and plant-based foods.[38] When drought in California raised public alarm over almond irrigation in 2014, Spain saw prices climb higher than ever. Farmers

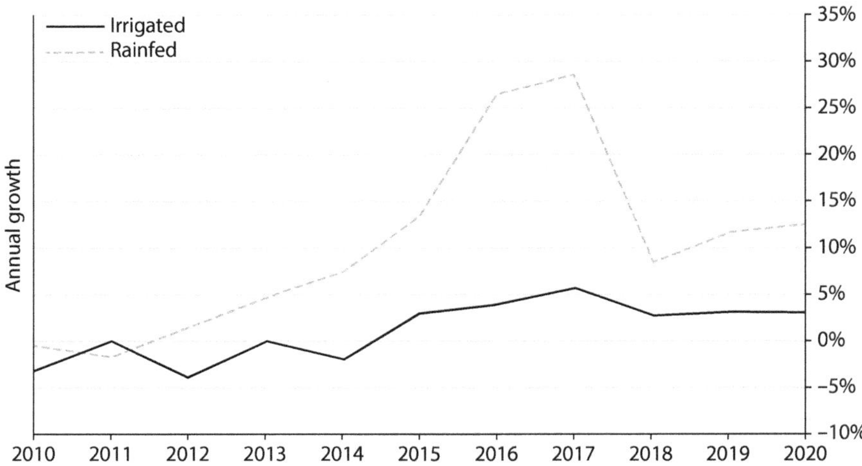

FIGURE 8. Spain's annual growth in areas planted with almonds. Data sourced from annual statistics of the Ministerio de Agricultura, Pesca y Alimentación. Analysis and figure by author.

planting new almond seedlings viewed this pattern as an opportunity. Demand had been growing steadily, and drought was sure to hit California again and send prices climbing. In yet another illustration of the market's ecological irrationality, California's drought is partially responsible for a planting boom in equally water-stressed Spain, where droughts are no less likely or severe. But so long as the timing of water limitations across geographies was right, there was money to be made.

Spain's total almond area had been steadily declining until 2015 when it abruptly changed course. Most of this expansion reflects new intensively irrigated plantings, which had seen an uptick beginning in 2013 (nearly 5 percent more than the year prior) and then surged, growing 26.7 percent and 28.6 percent annually in 2016 and 2017, with double-digit growth persisting through 2021.[39] While irrigated land had hovered around 7 percent of total almond-producing farms for over a decade prior, by 2020 it had reached nearly 20 percent. Rainfed orchards have also seen some growth, expanding more than 3 percent in 2015, peaking at nearly 6 percent growth in 2017, and sustaining over 3 percent growth ever since. Given the enormous yields from irrigated orchards, however, they quickly came to dominate. By 2021, just eight years after the beginning of the boom, irrigated almonds accounted for more than half of all production.

In this process, three major spatial shifts took place. First and most significantly, irrigated regions already suffering from the extractive practices of industrial agriculture saw a sudden almond planting spree. Farmers in the plains of Albacete (Castilla–La Mancha), Zaragoza (Aragón), and Sevilla (western Andalusia), faced with painfully low grain prices due to overproduction across Europe and with sluggish markets for wine grapes and olive oil, took a leap into almonds.[40] Much like California's Central Valley, Spain's irrigated valleys have become increasingly reliant on groundwater to sustain intensive production models as droughts rise in frequency and severity. Aquifer withdrawals are notoriously hard to regulate and, despite strict groundwater laws on the books since 1985, Spain struggles to restrict illegal wells. In 2006, the World Wildlife Fund issued a widely circulated report claiming 45 percent of all groundwater withdrawals were illegal,[41] a figure that rose to 70 percent in a 2017 follow-up report by Greenpeace.[42] Given the well-known responsiveness of almond trees to irrigation, high almond prices provide strong incentive to pump water that can net a few more kilos per hectare. Regional politicians have also tended to approve well construction and pumping under "emergency" drought conditions,[43] a dynamic likely to intensify with the economic losses posed by cutting flows to high-input permanent crops.

The almond boom drew production to some of Spain's most imperiled aquifers and ecosystems. Albacete sits atop an aquifer suffering severe groundwater over-exploitation, where overpumping has dropped the water table such that it drains, rather than feeds, the Jucar River.[44] Farmers in Sevilla have faced fierce backlash for groundwater extraction that depletes the waters of Doñana National Park, a UNESCO World Heritage Site, and endangers its sensitive wetland ecology.[45] Albacete and Sevilla, as well as Zaragoza's Ebro Valley, are plagued by high nitrate levels from heavy fertilizer applications.[46] While much of Europe has seen declines in nitrate pollution, Spain's levels have been increasing due to the concentration of high-value horticultural production.[47] How almond production might impact these dynamics will take time to assess. Intensive almonds are not necessarily more harmful than other industrially managed crops, but a spike in prices has rapidly folded them into sites rife with existing industrial agricultural hazards.

A second significant extension of almond acreage came in newly surface-irrigated zones, both within and beyond Spain's borders. The Segarra-Garrigues Canal in Catalonia prompted almond plantings by long-time farmers and real estate investors alike. An established mid-size farmer I met explained that given the costs associated with installing new irrigation infrastructure after the arrival of the much anticipated canal water, almonds simply made the most sense. The Ocean Almond company (mentioned in chapter 2), owned by construction and hotel firm OD Group, had set up shop nearby. In 2018 the company had purchased or leased over a thousand hectares surrounding the small town of Alcanó, where many elderly farmers signed over their land in exchange for the promise of stable income. The rainfed landscape transformed into an intensive almond plantation seemingly overnight. Similarly, the construction of the Alqueva Dam in neighboring Portugal had opened up an irrigated frontier, with almond plantings expanding at a pace of 15–20 percent per year.[48] The self-proclaimed "almond doctor" of California fame shifted his career there in 2018, moved much of his blog content behind a paywall, and declined my request for an interview. One Spanish agronomic adviser whom I interviewed was building his career consulting on new almond plantings. He eagerly recounted the ambitions of his newest clients to me, including Spanish companies with land in Morocco and Algeria. A few weeks after our interview, he sent me a photo on WhatsApp that looked like the Sahara Desert with twigs poking out of the sand. The Moroccan state has encouraged the development of newly irrigated almond plantings by Spanish firms as a form of foreign direct investment. Often entirely managed by Spanish staff, there is arguably little evidence of benefits to local residents.[49] As illustrated in the California case, the most arid of the irrigated zones may hold significant advantages when water can be procured.

The third geographic shift was an expansion of rainfed orchards. In 2019, the Ministry of Agriculture, Fisheries and Food estimated that 20 percent of land devoted to rainfed almonds had been planted within the five years prior. This growth took place in areas with long histories of rainfed production, like Eastern Andalusia, Murcia, and the inland grain-dominant region of Castilla–La Mancha. Unlike generations prior however, these almonds are not at the margins of other activities, they are the main attraction: uniform varieties in easily mechanized straight rows. The surge in rainfed almond plantings was a sign of hope and opportunity for some of the most "disfavored" regions, as I will illustrate. The newfound profitability of almonds offered a socioecological vision for rural revitalization. Yet like any boom, farmers knew the prices couldn't stay high forever. They hoped the crash wouldn't be too steep or come too soon.

A rainfed monoculture in a global market runs severe risks. Almond production is well known to swing dramatically year to year in response to variable rainfall and poorly timed frosts. In a typical agricultural market, one expects prices to rise when supply falls, cushioning the blow a bit for producers in years when yields are in a slump. (Of course, diversification reduces risks even further). Intensively irrigated production systems, however, exist in a world effectively detached from the weather. Given California's overwhelming dominance of the global market, hovering around 85 percent of all production, Spanish production has no effect on prices. When inclement weather dropped yields between 33 and 70 percent in 2022, prices stayed persistently low, falling from 5.56 euros per kilogram in August 2019 to 3.33 euros in August of 2020, 3.37 in August 2021, and 3.58 in August 2022.[50] Some growers reportedly chose not to harvest their crop because the cost of equipment rental and gasoline wasn't worth the effort. It is nearly impossible for rainfed production to compete with irrigation on market terms (though organic certification offers one strategy).

The problem for rainfed growers extends beyond susceptibility to weather and prices. The rise of intensive production also fundamentally changes the expectations of almond buyers. Industrial food production and snack packaging favors consistency. Rainfed almonds naturally vary in size and appearance with differences in rainfall patterns. Trees cultivated outside the standards of modernist plantations also reflect an enormously diverse gene pool, as farmers over generations have selected their preferred trees without systematic varietal distinctions. Rainfed orchards also carry the possibility of a few bitter nuts in the mix, due to preferences for directly seeded (rather than nursery bought) rootstock, which internal national buyers will not accept. Even the industry producing turrón, an almond confection with a protected geographical indication, now sources more than half of its almonds from California.[51] Importantly, this is all about the pursuit of industrial consistency, as prices for Spanish almonds are almost always lower. Rainfed production has very low input costs, but as one agricultural union put it, "super-intensive production is crushing traditional almonds."[52] In 2021, Spain's irrigated orchards produced more than 50 percent of the country's total crop for the first time, and rainfed growers feared intensive farms might rapidly dominate the market. Considering that almond growing is often not a full-time enterprise for rainfed growers, poor market conditions may push them out of agriculture altogether.

The potential for Spain's rainfed almonds to be edged out by their irrigated peers is a problem that extends well beyond the farmers themselves. The country faces an existential predicament surrounding rural depopulation, which combines questions of national identity with major concerns about the ecological ramifications of unmanaged spaces. The almond tree's newfound economic power has placed it at the center of efforts to revalorize rural livelihoods in rainfed regions through organic and agroecological methods. These market niches for high-value products have limits, but they might just be able to serve as anchors for a more stable future.

The Disappearing Margins

In the Spanish government's annual "Survey on Land Area and Yields of Crops," a new row appears in 2010 in the long list of fruits, vegetables, and grains: "abandoned almonds."[53] No other crop has a category dedicated to tracking landscapes of abandonment. Only almonds. Overgrown almond plantings are in many ways emblematic of Spain's rural exodus. Until the recent boom, they were a crop of last resort, a way to make productive use of the most marginal land with minimal time and resources. As economic conditions changed, fields of rainfed wine grapes, olive trees, or cereals would likely be transitioned to another agricultural use, but almonds are at the limit of agricultural viability. They are effectively the end of the line. When it is no longer worth it to maintain almonds, there is not much else one can do.

The hollowed-out villages and unmanaged fields marking a depopulated countryside have been given the ominous label "España Vacía," meaning Empty Spain.[54] This demographic shift has complex drivers common to industrializing economies. Low farm prices along with a promise of better wages and social mobility drew rural dwellers to urban centers. Yet the pace of change was incredibly swift in Spain.[55] It was not until the second half of the twentieth century that rural populations began a marked decline, dropping by about 40 percent by the new millennium.[56] The "Spanish miracle" of the 1960s marked an abrupt policy change in which dictator Francisco Franco's regime shifted from nationalist insistence on self-provisioning to a more liberalized economy.[57] The minister of agriculture at the time famously advanced the industrialization of the countryside by advocating for *"menos agricultores y mejor agricultura"* (fewer farmers and better farming).[58] Many rural residents squeezed out in this process emigrated abroad to send back remittances for their families. Following Spain's transition to democracy, entry into the European Union in 1986 coupled a flush of wealth in urban centers with further removal of trade protections for agriculture, which drove many more Spanish farms out of business.[59] In the early 2000s, debt-financed construction served as the major engine of economic growth, a pattern which drew low-wage workers away from rural zones while seeding a devastating economic crisis when the bubble eventually burst.[60] Many farmers in towns I visited noted that agriculture served as a fallback occupation whenever construction and tourism money ran dry, yet over the long run rural population decline has stubbornly persisted.

Mounting public concern with rural depopulation is not only a matter of equity or national identity,[61] it is also tightly bound up with questions of environmental protection. So much so that in 2020, the Spanish government created a new Ministerio para la Transición Ecológica y el Reto Demográfico (Ministry for Ecological Transition and Demographic Challenge). The ministry's combination of responsibilities reflects the centrality of rural population decline to visions for confronting climate change and biodiversity decline.[62] In clear contrast to American ideals of nature as devoid of human presence, many environmental advocates in Spain see stable rural communities as essential for a more ecologically sound future. Wildfires are among the most urgent threats posed by hotter and drier weather in the southern Mediterranean and are themselves massive sources of carbon emissions. As rural residents have moved to cities, the dense vegetation of unmanaged landscapes left behind has become fuel for increasingly uncontrollable blazes. In a 2022 report, the World Wildlife Fund argues that wildfire, as both a product of and contributor to climate change, "can only be mitigated by slowing depopulation, [and] encouraging sustainable rural development and traditional extensive land uses."[63] This is a particularly striking statement coming from an organization typically associated with shielding endangered species through fortress-like parks that heavily restrict Indigenous land uses.[64] Other groups like Commonland (discussed in more detail in the section to follow) see rural residents as crucial to Spain's fight against desertification, a broad and scientifically contested term that describes declining soil productivity.[65] Low-input agriculture, they argue, is essential for rebuilding soil quality. Current federal proposals directly link the fight against depopulation with a fight for a greener future, envisioning rural communities enlivened by renewable energy industries, conservation activities, sustainable agriculture, and ecotourism.[66] Certainly some ecologists view land abandonment as a boon for biodiversity,[67] yet these assessments are almost always tempered with an appreciation for the ecological benefits of cultural landscapes.[68]

Among farmers I spoke with, the absence of human presence was directly linked with a rise in undesired pest species. In Aragón, along the base of the Pyrenees mountains, several farmers I met blamed depopulation for a plague of wild boar. With enough people working in their fields, the boar would mostly stay away. And when shepherds still ran sheep and goats in the hills, there was much less food for the boar, keeping their population levels low. Now boar trampled through crops unabashedly, gobbling up grains, digging up roots, and breaking tree branches. Traffic accidents had become a nuisance, and some feared disease spread.[69] One pair of farmers showed me to the modest hut they used for hunting, photos lining the walls. "Look, here I'm going to give you an explanation. When we hunted these, our grandparents said it was a massive hunt . . . fourteen or fifteen. That's why we took the photo." "This year," he looked over at his friend, "How many did we hunt in one day?" "Sixty-five or sixty-four on some hunts," he replied as if freshly surprised by the sheer volume of carcasses. They blamed not only the decline in rural residents but also the shift toward richly fertilized maize in place of rainfed olive and almond trees, which provide much less fodder. Another almond grower living in a nearby town explained that a generation ago there were hardly any boar simply because "there had been human presence, and the presence of extensive herding, and the boar didn't have much space. . . . Now you go across wide stretches and you don't see a soul." Wild boar have a place in Iberia's ecology, but their rapid proliferation epitomized for many rural residents a troubling imbalance wrought by absence.

Spain's rural exodus has complex drivers unlikely to abate anytime soon, and certainly the fate of a single crop like almonds is but a drop in the proverbial bucket. Yet the rainfed almond tree offers an opportunity to consider marginality not as a defect but as a strategic asset. It is no economic miracle crop, as briefly believed in the 1970s, nor will it repopulate the countryside. Rainfed almonds are a secondary or peripheral activity for many households. For some, they are the only agricultural activity that keeps people connected to the land. While many agronomists bemoan what they see as a missed opportunity, given such minimal maintenance, sporadic engagements are in many cases a remaining thread in what might otherwise be a resignation from agriculture all together.

Some rural residents I met lamented that almonds have become the favored crop of *domingueros*, a term for urban dwellers who come back to farm only on Sundays. For these families, a low-maintenance crop like almonds provides a sense of continued connection to land and a bit of extra income from subsidies and sales. For many, these small plots were the pride and joy of their grandparents, purchased after many years of diligent saving after toiling for the benefit of privileged landowners. Walking the fields is akin to flipping through a photo album, certain trees or sites prompting embodied memories. Here the speed of Spain's agrarian change is vividly displayed. A retiree in rural Granada showed me around his modest plot, pointing to the remnants of a stone structure where he remembers tying up a donkey as a child and sorting wheat to be threshed by hand. Tending to almond trees, along with woodworking, kept him active. While policymakers seeking to boost the number of registered residents in rural zones might not see weekend farmers favorably, low maintenance crops like almonds foster the continued tending of agrarian landscapes.

European Union subsidy schemes have increasingly recognized this, paying rural landholders for crops with a "high nature value" or for ecologically mindful practices. The most recent 2023–27 package specifically provides support for rainfed almonds (along with hazelnuts and carob) in regions at risk of desertification, as designated by rainfall and slope. Almonds are noted as being particularly well-suited to erosion prevention and the impacts of climate change, and there are additional subsidies where vegetative cover or mulch is applied.[70] Farmers who integrate almonds with other crops, however, are likely to face challenges receiving payments which have historically been predicated on monoculture land management categories. Time will tell if these subsidies, along with national demographic policies, are sufficient to achieve their intended ecological effect and what kinds of unintended consequences may result. Technocratic economic policies have also been critiqued for missing the mark on rural development because they "consider territory in a linear and mechanical way rather than as a complex system of interdependent social, cultural, institutional, affective and political relations involving people" first and foremost.[71] Without examining the desires and experiences of the people living within these dynamic worlds, critics argue, efforts to stabilize rural communities by means of agricultural economics alone are doomed to fall short.

Hopes for an "Almendrehesa"

A more holistic approach for fostering a thriving rural landscape appears to be emerging in the nonprofit sector, where one group concerned with land degradation has mobilized around the unique qualities of almond agroecosystems. In the high plains of eastern Andalusia and western Murcia, a project called Almendrehesa takes its name from the words *almendra* and *dehesa*, a rainfed cultural landscape of integrated livestock, oak trees, and grains historically common to southern Spain and Portugal. As the name suggests, the vision is to protect and revitalize rural landscapes through a highly diversified agroecology centered on almond trees with an understory of legumes, grains, and aromatic herbs integrated with sheep grazing and beekeeping. Natural vegetation strips provide habitat for wild pollinators and natural enemies of crop pests. Soil organic matter is gradually built up by reducing tillage, adding compost, spreading manures, and processing branches removed during pruning into a fine mulch. Ponds, swales, and restored terraces help slow rainwater so it can percolate into the soil. Farmers and other local businesses create value-added products based on ecological restoration to sell at a premium. These products and their origins become the basis for culinary and agricultural tourism. It is a compelling vision that has grown from an initial cohort of 21 members in 2017 to over 460 in 2022.

The Almendrehesa project seeks to demonstrate the synergies among elements within an agroecological whole that go beyond the prescriptions of subsidies or input restrictions of organics. I met with one of the founding members, a woman who was raising a thousand sheep on sixty acres (twenty-four hectares) of grains and legumes grown interspersed with almond trees. She worried that high prices or subsidies would create a monoculture of almonds. "What the land, the countryside, the environment really needs is a system we can mosaic, which is to say, where there is a bit of everything, like there always has been, and we're very concerned that . . . it will all only be almonds." This revalorization of diversified agricultural spaces as essential to environmental repair echoed those of another founding member of the group, a long-time farm consultant and supplier of organic inputs in the region. "We are contributing to the conservation of natural space, of biodiversity . . . contributing in a very important way to the agrarian countryside, what we call *paisaje agrario*, which really is as important as the natural landscape, if we can even call it natural, because in the end everything is touched by humankind in one way or another."

Building new knowledge networks was crucial, he went on to explain, and a highly individual process. "We've seen this land go from subsistence agriculture to subsidy agriculture. There has been money, but money doesn't change mentality. I think in a way what it does is distort certain social values, and [regenerative farming] was not a question of money, it was a question of changing mentality." Many farmers are unfamiliar with the complexities of diversified systems, and they typically follow whatever the subsidies require. Almendrehesa has been providing ongoing trainings, as the director explained, because many farmers have become accustomed to doing things a certain way, for example tilling the soil in excess to make a field look clean. Research and farmer demonstration plots have helped bolster confidence in practices like cover cropping that many growers were worried might draw too much water and nutrients away from the primary crop.[72] Formal education is helpful but ultimately limited when farming cannot be reduced to a recipe, the organics consultant explained. "Classes and everything help you to have an orientation, but if you really want to regenerate, restore, recuperate, you have to recuperate with the knowledge of the land you walk on and know. If you don't know it, you won't be able. . . . This only happens from within, from within the knowledge of each farmer." This kind of observational, embodied knowledge is at odds with the scalability at the core of industrial agronomic training.

Participants in the local organization AlVelAl,[73] of which Almendrehesa is one component, also present a social vision for a fulfilling rural life that they hope can attract a younger generation. Many young people left urban centers to return to their rural roots in the wake of the 2008 economic crisis, and AlVelAl's leaders hoped that the satisfaction of collective land restoration work would make it hard to leave again. Ecological restoration is not possible, the organization's leaders maintain, without a path toward a dignified life (*una vida digna*).[74] They aim to foster meaningful employment in agriculture and agrotourism along with events that celebrate artistic and cultural life. The organization's slogan centers the hope that agroecological practice can slow depopulation: "regenerating soils, rooting people."

Almendrehesa's longevity, however, is uncertain. The impetus for the project came from the Dutch nonprofit Commonland, which began seeking out regions around the world at risk of desertification. Its financial viability at present rests on donor funding, with hopes that by the end of the program's twenty-year commitment, it can attain financial independence. Commonland's founder believes deeply that the right combination of entrepreneurship and ecological restoration can provide a self-reinforcing cycle of prosperity, but the evidence on this point is far less clear. Premium "beyond organic" labels, along with agrotourism, have a limited market and can be sensitive to economic shocks.[75] Many Spanish towns pursuing rural tourism with an "if we build it, they will come" attitude have failed to see the desired results,[76] particularly when they are beyond a daytrip for beachgoers.[77] And even the most ecologically mindful tourism presents an environmental contradiction, considering the material impacts of travel.[78]

The private sector alone is unlikely to sustain this kind of project. Using an ecological economics approach, researchers evaluating the Almendrehesa polyculture in comparison with monocultural organic almonds and conventional almonds concluded that even if all "externalities" (such as erosion control and carbon sequestration) could be exhaustively accounted for, public investment would still be necessary.[79] There is a business case to be made for organic monoculture, which is thriving. Spain is the world leader in organic almonds, producing more than seven times that of California in 2021, largely because the rainfed low-input system makes good economic sense.[80] But consumer demand for boutique products and experiences alone cannot be counted on to holistically transform rural landscapes. While organic almond prices have remained relatively more stable, the premiums relative to conventional production may shrink as large producers enter the picture, putting smaller, more diversified, and more topographically complex operations at a disadvantage. The alluring vision of green capitalism remains tenuous because the problem lies not in internalizing "externalities" but in the market itself, which is simply a blunt and inadequate tool for achieving social aims.[81]

PROFITABILITY DRIVES TOWARD PLACES OF PRECARITY

Notions of landscape suitability, while seemingly an inherent plant characteristic, are intimately tied to political economic processes and modernist ideologies of more-than-human mastery. In both California and Spain, the rising profitability of almonds has shifted their geographies, introducing new ecological precarities. In California, almonds have gradually expanded from the northern and central regions of the Central Valley into its southern counties. Southward expansion has introduced new pest pressures, as insect reproduction is not stunted by freezing winters. Hot, dry weather along with increasingly saline soils demands higher irrigation rates. Yet orchards in the valley's southern stretches have become icons of agronomic success. They are considered more "efficient" by growers, agronomists, and industry leaders because warm, dry conditions when paired with agrichemicals significantly boost production. In Spain, the almond boom is shifting production inland, away from upland slopes and toward the irrigated plains. This new geography is transitioning almonds into some of the country's most overexploited and contaminated aquifers. Such landscapes are championed as solving at long last the irregularities of rainfed production with reason.

The geographic expansion of the almond boom pushes production into increasingly precarious terrain. This poses hazards both for environmental integrity and long-term economic viability. These spatial reorderings are both highly efficient and precarious, revealing the distinctly place-based limitation of calculating production efficiency as the prime metric of sustainability. Efficiency in terms of yield per land area or "crop per drop" is imagined to exist outside of place and time, but the materiality of agricultural land tells a different story. Place matters profoundly, and accumulating ecological burdens will eventually take their toll. Almonds at the margins are inefficient by definition, but they offer an alternative spatial logic of production. There the specificities of a place and its people come first.

5

Conjuncture

Rooting Agricultural Knowledges in Place

The ills of industrial agriculture are well known. Pesticides allow dense mono-cultures to thrive uninhibited by the organisms naturally drawn to their abundance, yet they do so for a limited time and at substantial human and ecological cost. Pollinators suffer the compounding stressors of pesticide sprays, parasites, and a landscape incompatible with authentic nourishment. Rivers have been rerouted, aquifers depleted and physically deformed, and waterways of all kinds contaminated with nitrates and other toxins. Soils have been eroded and loaded with salts demanding yet more water to keep agriculture viable. Wells run dry during droughts, straining farmers and disproportionately impacting the homes of marginalized rural communities. These phenomena are central features of the almond boom in California and more recently Spain, but they are not unique to almonds. Anywhere that extractive agriculture has taken root suffers similar burdens, particularly the arid regions where industrial efficiencies have been optimized.

I will not be concluding this book with a set of recommendations for how almond agriculture could be done differently, though this work is undoubtedly important. Projects to grow almonds without such ills are active in a range of contexts and scales. The holistic vision of Almendrehesa to create an agroecologically diverse landscape has some parallels among a few enthusiastic farmers in California's Capay Valley. More incrementalist approaches in California aiming to reduce pesticide toxicities or nitrate losses have parallels among Spanish researchers and farmers. Each project has its merits and limitations. But I will not be ending with such examples for several reasons.

First, because doing so might give the impression that the fundamental problems with intensive almond production are technical in nature and thus can be resolved by technical means. What I have argued throughout each chapter is just the opposite. The problems laid bare by the almond boom are political economic in nature, and any potential remedy must address them as such. A second, related reason not to end on improving almond agriculture is that doing so would risk taking current levels of almond consumption for granted as a static nutritional requirement to "feed the world" or even as an ethical failure of thoughtless eaters. Instead, I am insisting here that record-high global almond consumption is a byproduct of intensification, much like the shrinking aquifers or precarious pollinators previously described. People are eating more almonds than ever because California growers have organized and advertised, wielding a powerfully persuasive marketing machine to stay ahead of a looming overproduction price crash. A third reason to resist ending with an optimistic story specific to almonds is because it might give the impression that this book is about almonds, how harmful they are, and how they can be reformed. What almonds are today, another crop will be tomorrow. This book project has never, at its core, been about almonds.

This book is about questioning the claims about what agricultural plants need that are so often taken at face value. How much water does it require? How much does it depend on managed pollination? Where is it best suited to grow? These appear at first glance to be biophysical questions with answers that can be deduced from botanical traits. But upon closer inspection, even something as simple as measuring evapotranspiration rates is profoundly political. The assumptions built into the equation, the state-sponsored network of weather stations atop twenty-acre lawns required to implement it, and the capacity for corporate farms to invest in research at immense scales are all baked into the number that results. The prominent place honeybees play in contemporary food production is often portrayed as if it were a function of their biological job as pollinators. Without masses of managed hives, the implication follows, we might starve. Yet the relationship between beekeepers and farmers is much more of an economic codependence, each using the other to cope with its own crises of profitability. Both parties have depended on one another to survive the industrialization of the landscape: intensifying orchards became incapable of sustaining sufficient populations of local pollinators while mobile beekeepers leaned on monocrop farming to boost honey yields. These are relations bound up with capitalist overproduction pressures that no botanical study can reveal. Yet somehow the language of the agronomic sciences has come to naturalize extractive relations among plants, water flows, and honeybees as if they were matters of physical necessity. I hope that readers come away from this book with the message that no statement about what plants need is as static or universal as it appears. Each presumed necessity might be better followed by this question: To what extent does this biological "requirement" reflect a political economic configuration, and how did it come to be so?

Yet the material characteristics of agroecosystems are not a neutral backdrop against which capitalist processes play out. My aim has been to show how materiality and meaning are deeply intertwined. A given almond tree's root structure, shell hardness, or sensitivity to frost is not merely a constraint on capital. Each characteristic embodies layers of historical circumstances that both intentionally and unintentionally prioritized certain ways of being over others. In the case of early breeding efforts, California's economic dependence on the paper-shell Nonpareil almond is inseparable from the racial capitalist logics of its creator. The thinner the shell, the more hope for a profitable industry that would replace Chinese residents with white settlers. Ultimately this vulnerable shell type resulted in a deepening addiction to synthetic insecticides and their poisoning of lives and landscapes. When it comes to the almond boom's shifting geographies of production, the combination of irrigation, aridity, and flat mechanizable landscapes has drawn orchards in both California and Spain toward hotspots of preexisting precarity. The materiality of the almond tree and its responsiveness to water, dry air, and a warm extended growing season are folded into the landscapes sculpted by prior iterations of extractive agrocapitalist enterprise. The material is always simultaneously political and vice versa.

Comparative research is a marvelously messy undertaking. By taking a diffractive approach, I have attempted to read the cases of California and Spain through one another, looking at the kinds of patterns that form when they are forced into the same frame. I find this form of relational comparison to be not only analytically fruitful but also authentic to how I experienced the research process. The analysis I offer in this book is partial and situated, concerned not with creating clear bounded objects to be contrasted but instead allowing the most potent stories in each place to speak to one another. Treating these two cases in relation makes it easier to denaturalize one particular way of knowing, opening alternate paths and possibilities. One element that attracted me to this approach is the way it tempers the propensity for capitalist processes to appear all-encompassing. Like California, Spain's agricultural economy has been dominated by capitalist configurations since at least the mid-nineteenth century, yet the persistence of rainfed almond landscapes reflect a relationship to land that quietly resists wholesale commodification. By understanding almond trees as a rustic, resilient, and pragmatic component of a polyvalent rural life, rainfed farmers often reject productivity increases if they might compromise longevity. For the most part, these are not ideologically motivated individuals pursuing an agroecological vision, they are simply doing what makes sense within a multifunctional landscape. The way formal agronomic knowledge in Spain had focused almost entirely on plant breeding until the boom of the 2010s reveals how tightly knowledge production is tied to circuits of capital. Only when almonds caught the eye of highly capitalized irrigated growers and farmland investors did the research investments in pests, plant nutrition, and other productivity enhancements follow. And once the switch had been flipped, the codified agronomic knowledge from California supercharged Spain's production boom. Like following a recipe, new irrigated plantings reached astounding levels of intensification within just a few years that had taken decades of research and considerable convincing for California growers to adopt.

The almond paradox makes clear that agricultural knowledges are undeniably contextual, forever bound up with the people and places of their development, even as they travel across time and space. A given logic, whether prioritizing industrial efficiency or rural resiliency, also carries very place-specific impacts. It doesn't matter how much "crop per drop" you can squeeze out of an acre of almonds if the aquifer below that acre is drained and contaminated at an accelerating rate. And yet modernist agronomic thinking treats the farm as a factory-like bubble of inputs and outputs. It lacks a mechanism for incorporating the broader ecological context of the farm, let alone the values and desires of communities where farming takes place. The goal is extremely narrow: optimizing productivity and profitability. Ironically, the more almonds one churns out from a given plot of land or volume of water, the worse the problem becomes. Efficiency simply feeds into the cycle of overproduction as yields surge and growers panic over the ability for the market to absorb such abundance. It is clear that such a placeless paradigm generates rushes of profits for a given window of time but ultimately renders rural lives of all kinds profoundly precarious.

So where do we go from here? Given what the almond paradox reveals, I see several lessons learned. One is that efficiency is meaningless as an environmental indicator without place-specific goals. Anywhere the word *efficiency* is invoked in the name of sustainability, we should be asking, What is the ambition for that particular place, and how (if at all) is a given form of efficiency moving toward it? For example, take the argument that almond production has become more benign by increasing production relative to inputs like water, fertilizer, or agrichemicals. On the surface it sounds like an improvement. But if you couple these statements with measurements of water levels, contamination, and pesticide applications in prime production regions, the trendline looks quite different. For those living alongside intensive orchards burdened by falling water tables and toxic exposures, a mounting volume of almonds pouring out each year means very little. Efficiencies of certain kinds could very well support better futures for rural communities. Without a connection to place-specific outcomes, however, efficiency claims are worse than merely misleading. They are complicit in compounding existing problems and legitimating people and places as destined for sacrifice.

A second, related takeaway is that prevailing forms of agronomic knowledge that obscure or downplay ecosocial contexts are hazardous. An industrial model of inputs and outputs is privileged when optimizing for profits. But such crude economic utilitarianism clearly undermines long-term rural livability, especially for those people already treated as disposable. This is not a coincidence, it is the second contradiction of capital. As a system, racial capitalism exploits social differences while undermining the material basis of its own existence.[1] A scientific process guided by capital generates exactly what one would expect. Instead, imagine if failing to consider the potential social and ecological ramifications of farm management strategies was considered intolerable. Perhaps it would be flagged during peer review, much like other research ethics protocols. Imagine if the resources the state currently dedicates to purportedly placeless productivist agronomy were put toward place-based projects, like the experiments engaged by Almendrehesa and others, to craft a multifunctional landscape suited to the local ecology and community. Imagine if agricultural knowledge production was explicitly antiextractive. Much as the movement for antiracism requires moving from a passive denouncement of injustice to an active engagement in its dismantling, agricultural researchers might work to make naturalized extraction unviable. As the famous ethical principle attributed to Hippocrates states: first, do no harm.

All of this puts quite a lot of pressure on knowledge purveyors, academic or otherwise. When public institutions are tasked with prioritizing farm profitability, often in partnership with agribusiness or corporate farming enterprises, a productivist outcome is not surprising. California's agricultural extension system has shifted some resources toward sustainability aims, and Europe's latest Common Agricultural Program (CAP) takes significant steps to prioritize environmental objectives. Both have met substantial resistance in the farming community. Many of the California growers I interviewed resented the university's broadening social goals as an irrational imposition coming from urban centers. Agronomists I met often described choosing their line of work because helping farmers make a better living felt personally rewarding. I could understand their hesitance to give up some of the industry accolades and the satisfaction of direct impact. In Europe, policy announcements for greening initiatives were met with literal roadblocks. In January 2024, farmers across several European capitals, including Madrid, clogged roadways with tractors for weeks to protest the costs imposed by the new CAP. EU leadership quickly conceded to scrapping one of its central goals: to halve pesticide use by 2030.[2] Research and policies that are explicitly antiextractive will be an uphill battle with significant structural barriers.

The crux of the matter rests on solving the ecological riddle of modern agricultural markets. Showing that agroecologically vibrant landscapes are possible and desirable in a confined setting will not necessarily make them economically viable or able to replace the dominant extractive practices. Many alternative agriculture movements around the globe have been limited by an "if we build it they will come" mentality, constantly struggling against the constraints of the market. Products relying on a price premium can, by definition, only be purchased by a segment of the population, and a progressive decline in real wages doesn't bode well for this strategy. Fundamentally, a better future requires a political economic system that favors agricultural forms serving societal needs rather than squeezing farmers between high costs and low prices such that productivism is the only path. Markets are poor arbiters of more-than-human well-being. The technological treadmill of capitalist agriculture sends half of all US-grown corn into gas tanks and generates mountains of almonds that marketing teams scramble to sell around the globe. A neoliberal fixation on educating consumers to make certain food choices will do little to reign it in.[3] Supply management policies that can slow the treadmill and stabilize incomes for farmers are an essential baseline for changing course.[4] They must be coupled with a political resolve to refuse compromising human and ecological health for the sake of private profit. The real power players in agriculture are not the farmers but the businesses selling the inputs and buying the outputs, and the treadmill of overproduction serves them well.[5] Given the state's historic role in greasing the wheels of racial capitalism, repurposing its mechanisms toward more livable and equitable rural futures will require sustained struggle, to say the least.[6]

One exceedingly modest suggestion is to avoid eliminating the agroecosystems that have managed to evade extractivism. Rainfed almonds in Spain are good to think with because they illustrate the importance of the margins, both economically and ecologically. According to industrial logics, margins are destined for erasure. Crops should be planted fencerow to fencerow in the places where maximum yields can be achieved. Smaller, part-time, and less capitalized farmers are destined to either get big or get out. Irrigated valleys and plains with prime agricultural soils become highly productive sacrifice zones while more diverse terrains lose their active agricultural engagements. Yet rainfed almonds illustrate the opposite orientation. They have long been relied on for their complementarity not just their productivity. For their ecological fit, not just their annual output. For their ability to endure not their ability to extract. This has as much to do with the almond's unique physiology as it does its historic place at the blurry boundaries of capitalist enterprise.

Perhaps a bit of low-hanging fruit in the political struggle for holistic agricultural reform is to fortify place-based efficiency where it already exists. As the classic Joni Mitchell song goes, "you don't know what you've got till it's gone." Rainfed almonds growers, like many peasant producers around the world, are making the best of the available resources while minimizing their expenses and efforts. Often almonds are a minor player in a diverse landscape or among diverse income sources. Rather than placeless industrial efficiency, they exemplify a place-based agroecological efficiency. Such pragmatism is born of necessity rather than ideology, and these landscapes carry their own complex legacies. My aim here is not to romanticize rainfed almond production, or any other particular place or product. Instead, I wonder how we might use such agroecological thrift as a kind of a minimum basic requirement for preventing future policies from making things worse. How well does a given effort enable existing ecologically resilient cropping systems to thrive? How well does it protect the margins from the perils of the market? This is a painfully modest proposal. It does not in any way provide a remedy for the prevailing extractivist paradigm and the engine of overproduction that drives it. But you have to start somewhere. Rooting agricultural knowledge in place means exposing its origins, making its place-based impacts explicit, and holding on to the already existing antiextractive, place-based knowledges where we can find them. First, do no harm.

INTRODUCTION: NATURALIZED EXTRACTION
AND KNOWING OTHERWISE

1. Weiser, "Lucrative but Thirsty Almonds."

2. Palomino, "California's Thirsty Almonds."

3. Romero, *Economic Poisoning*; Gorman, *The Story of N.*

4. O'Connor, "Two Contradictions of Capitalism."

5. Pellow, "Critical Environmental Justice Studies."

6. Whyte, "Settler Colonialism, Ecology, and Environmental Injustice."

7. World Economic Forum and McKinsey & Company, *Innovation with a Purpose.*

8. United Nations Environment Programme, "Sustainable Consumption and Production Policies."

9. CropLife International, "Advancing Innovation in Agriculture."

10. Almond Board of California, *Facts About Water Use Among Almond Growers.*

11. Faunt et al., "Water Availability and Land Subsidence."

12. Allen, *The Almond People.*

13. Tucker, "The Future of the California Almond."

14. Reisman, "Superfood as Spatial Fix."

15. Stoll, *Fruits of Natural Advantage.*

16. Walker, *The Conquest of Bread.*

17. The diverse resident populations, known now by the groupings ascribed to them by their colonizers, included the Kumeyaay, Luiseño, Kumi'vit/Gabrieleño, Chumash, Salinan, Esselen, Ohlone/Costanoan, Coast Miwok Cahuilla, Yokuts, and Miwok, among others. For more, see Jackson and Castillo, *Indians, Franciscans, and Spanish Colonization.*

18. Dillon, "Civilizing Swamps in California."

19. Many almond plantings in the Paso Robles region were the result of land speculators dividing larger plots into ten-acre "orchard bungalows," which they sold to remote

investors. A group of investors later sued the company for misrepresenting the quality and profitability of the orchards.

20. Tucker, "Future of the California Almond," 16.

21. Tucker, 5.

22. Allen, *Almond People*, 91.

23. US General Accounting Office, "Role of Marketing Orders."

24. US General Accounting Office.

25. The "Got Milk?" advertising campaign, funded by the California Milk Processor Board, was funded using the same mechanism.

26. Reisman, "Superfood as Spatial Fix."

27. Reisman, "The Great Almond Debate."

28. Almond Board of California, *Almond Almanac 2022*.

29. US Department of Agriculture, Agricultural Marketing Service, "Almonds Grown in California."

30. Cochrane, *Development of American Agriculture*.

31. Philpott, "Lay Off the Almond Milk, You Ignorant Hipsters."

32. Hamblin, "Dark Side of Almond Use."

33. Associated Press, "How China's Taste for Almonds Is Sucking Drought-Stricken California Dry."

34. Agricultural Economic Insights, "U.S. Almond Production and Consumption Trends."

35. Grasselly, "Réflexions diverses sur l'évolution des objectifs d'amélioration de l'amandier."

36. Reisman, "Protecting Provenance, Abandoning Agriculture?"

37. A 1799 census indicates almond production was highly concentrated in Valenica, as well as eastern Andalusia and the Balearic Islands. Polo y Catalina, *Censo de la riqueza territorial é industrial de España en el año 1799*.

38. Vallés y Vallés, *El almendro*.

39. Joseph Harrison, "Agrarian History of Spain."

40. Tello et al., "Feudal Colonization to Agrarian Capitalism."

41. Murray et al., "Biocultural Heritages in Mallorca."

42. García Moreno, *El almendro*, 26.

43. Robledo, "La reforma agraria durante la Segunda República."

44. Jefatura del Estado (España), ley 157/1961, de 23 de diciembre, que prorroga la de 17 de julio de 1951 sobre repoblación de almendros, algarrobos, higueras, olivos y viñedos, Pub. L. No. BOE-A-1961-23887, § 311, 18359 (1961), https://www.boe.es/buscar/doc.php?id=BOE-A-1961-23887.

45. Ministerio de Agricultura, *El almendro*.

46. Riera, Lamich, and Calafell, *Cultivo del almendro*, 120.

47. Rubí, *El almendro*, 17.

48. Rivas Sanz and Fernández-Maroto, "Planning for Growth."

49. Ministerio de Agricultura, Pesca y Alimentación, Orden de 18 de julio de 1989 por la que se establece la normativa para la solicitud, control y pago de las ayudas para la mejora de la calidad y de la comercialización de los frutos de cáscara y algarrobas. Pub. L. No. BOE-A-1989-17178, 23090. 1989. https://www.boe.es/eli/es/o/1989/07/18/(2).

50. US Department of Agriculture, *2020 California Almond Acreage Report*.

51. The most recent data from Spain's Ministerio de Agricultura, Pesca y Alimentación reports production and land area through 2021.

52. Political ecology is a wide-ranging body of scholarship that examines how power relations—including capitalism, colonialism, patriarchy, and other forms of oppression—operate through environmental change. Sundberg, "Feminist Political Ecology."

53. Sayre, "Genesis, History, and Limits of Carrying Capacity."

54. Brockington, "Ecosystem Services and Fictitious Commodities."

55. Mitchell, "Fixing the Economy"; Heynen and Robbins, "Neoliberalization of Nature."

56. Brown, "When Food Regimes Become Hegemonic."

57. I distinguish here between feminist STS and science and technologies studies writ large, given the latter's tendency to sideline political economic relations where feminist scholars have held them front and center.

58. Haraway, "Situated Knowledges."

59. Latour, *The Pasteurization of France*; Kimura, *Radiation Brain Moms and Citizen Scientists*; Adams, *Glyphosate and the Swirl*; Suryanarayanan and Kleinman, *Vanishing Bees*.

60. Navon, "Reiterated Fact-Making."

61. Kloppenburg, *First the Seed*.

62. Fitzgerald, *Every Farm a Factory*.

63. Freidberg, *French Beans and Food Scares*.

64. Freidberg, *Fresh*.

65. Henke, *Cultivating Science, Harvesting Power*.

66. Guthman, *Wilted*.

67. Bronson, *Immaculate Conception of Data*.

68. Fox and Alldred, "New Materialist Social Inquiry"; Castree et al., "Mapping Posthumanism"; Ogden, Hall, and Tanita, "Animals, Plants, People, and Things"; Panelli, "More-than-Human Social Geographies."

69. Arboleda, "Revitalizing Science and Technology Studies."

70. Barad, "Posthumanist Performativity"; Haraway, *Manifestly Haraway*; Law, *Material Semiotics*.

71. Sundberg, "Decolonizing Posthumanist Geographies"; Todd, "An Indigenous Feminist's Take on the Ontological Turn"; Watts, "Indigenous Place-Thought and Agency"; Simpson, "Aboriginal Peoples and Knowledge."

72. Wynter, "Unsettling the Coloniality of Being/Power/Truth/Freedom."

73. Cochrane, *Development of American Agriculture*.

74. Stoll, *Fruits of Natural Advantage*.

75. Friedland, Barton, and Thomas, *Manufacturing Green Gold*; Mintz, *Sweetness and Power*; Wells, *Strawberry Fields*; Guthman, *Wilted*; DuPuis, *Nature's Perfect Food*; Beckert, *Empire of Cotton*; Soluri, *Banana Cultures*; Hetherington, *The Government of Beans*.

76. Gibson-Graham, *The End of Capitalism*. I also have Anna Tsing to thank for always reminding me of the patchiness of the Plantationocene.

77. Tsing, *Mushroom at the End of the World*.

78. Wynne, "Public Engagement as a Means of Restoring Public Trust in Science"; Corburn, *Street Science*.

79. Fairbairn, "Framing Transformation"; Holt-Giménez, *Campesino a Campesino*; Kinchy, *Seeds, Science, and Struggle*; Patricia Allen, *Together at the Table*.

80. California Department of Food and Agriculture, *California Agricultural Statistics Review 2022–2023*, 60, 100.

81. Federación española de asociaciones de productores exportadores de frutas, hortalizas, flores y plantas vivas, "Exportación/importación Españolas de frutas y hortalizas"; Fruit Logistica, *European Statistics Handbook 2023*, 3.

82. US Department of Agriculture, National Agricultural Statistics Service, Honey; EuroStat, "Beehive Numbers in EU."

83. Swyngedouw, *Liquid Power*.

84. Claire and Surprise, "Moving the Rain"; Swyngedouw, "Not a Drop of Water."

85. Jasechko et al., "Rapid Groundwater Decline."

86. Faunt et al., "Water Availability and Land Subsidence in the Central Valley."

87. Molle and Closas, "Why Is State-Centered Groundwater Governance Largely Ineffective?"

88. Custodio et al., "Sustainability of Intensive Groundwater Development."

89. Scott et al., "Irrigation Efficiency and Water-Policy Implications."

90. Benito and Pulgar, "Modernización agraria, modernización administrativa y Franquismo: El modelo educativo y administrativo del Servicio de Extensión Agraria (1955–1986)."

91. Romero Montero, "La ayuda norteamericana a España."

92. Díaz Geada et al., "Agricultural Extension Programmes in Post-War Europe."

93. California Almond Growers Exchange, *Shall the American Almond Industry Perish?*

94. Lowe, "Insufficient Difference."

95. Wallerstein, *The Modern World System*.

96. Blaut, *Colonizer's Model of the World*.

97. McMichael, "Incorporating Comparison Within a World-Historical Perspective."

98. Hart, "Relational Comparison Revisited."

99. Strathern, *Partial Connections*.

100. Tsing, "Strathern Beyond the Human."

101. Haraway, "Situated Knowledges."

102. Mol, *The Body Multiple*.

103. Barad, *Meeting the Universe Halfway*.

104. Haraway, *Modest_Witness@Second_Millennium. FemaleMan_Meets_OncoMouse*, 14.

105. Haraway and Goodeve, *How Like a Leaf*.

106. Barad, *Meeting the Universe Halfway*, 71. Cited in Van der Tuin, "Diffraction as a Methodology for Feminist Onto-Epistemology."

107. Reisman, "The Great Almond Debate."

108. I also happened upon a devastating outbreak of the bacterial disease *Xylella fastidiosa* in almond trees, which I explored in my paper "Plants, Pathogens, and the Politics of Care."

109. Evans and Jones, "The Walking Interview"; Mathews, "Landscapes and Throughscapes in Italian Forest Worlds."

110. Reisman, "Protecting Provenance, Abandoning Agriculture?"

111. Reisman, "Superfood as Spatial Fix."

112. Pistachios are the rising investment darling of California's Central Valley, in part due to the progressive salinization of soils (see chapter 3).

113. Guthman, "Neoliberalism and the Making of Food Politics."

1. MATTER: MEANING-MAKING IN A NUTSHELL

1. I use *cultivar* and *variety* interchangeably in keeping with the lexicon of my interlocutors. Among botanists, *variety* refers to plant subspecies with a distinctive suite of

characteristics occurring without intentional human intervention. Among commercial plant breeders and farmers, however, *variety* is more commonly used than *cultivar*.

2. Fitzgerald, *The Business of Breeding*.

3. Kloppenburg, *First the Seed*.

4. Legun, "Tiny Trees for Trendy Produce."

5. Guthman, *Wilted*.

6. Whatmore, *Hybrid Geographies*.

7. See Mathews, *Trees Are Shape Shifters*, for a methodological exploration of reading landscapes, particularly where woody tree species predominate.

8. My thinking here builds on material-semiotic theory, a school of thought associated with feminist science studies that rejects the dominant Western rendering of matter as passive. Many Indigenous theorists have long taken the active social role of more-than-human beings as a baseline assumption and further underscore the particularity and ethical weight of these relations. For an overview of these bodies of scholarship and their intersections and an analysis of how Indigenous scholarship remains marginalized, see Rosiek et al., "New Materialisms and Indigenous Theories."

9. For an exhaustive review, see Gradziel and Martínez-Gómez, "Almond Breeding"; Socias i Company and Gradziel, *Almonds*.

10. Lindsay, *Murder State*; Madley, *An American Genocide*.

11. Suarez, "A Legal Confiscation."

12. Balfour, "Riot, Regeneration and Reaction."

13. Wickson, *California Illustrated*, 87.

14. Norton, *Genocide in Northwestern California*.

15. Saxton, *The Indispensable Enemy*.

16. Wickson, *California Illustrated*, 88.

17. California State Board of Horticulture, *Biennial Report of the State Board of Horticulture*, 63.

18. Wey, "History of Chinese Americans in California."

19. Robinson, *Black Marxism*.

20. Wickson, *California Illustrated*.

21. Riley, "History of the Almond Industry in California."

22. Bitter rootstocks were and still are often preferred by many rainfed growers who value a deep taproot and more limited pest damage (see chapter 2).

23. References to the Nonpareil variety here represent the broader collection of Hatch's paper-shell almond varieties, of which the Nonpareil became most popular, including the IXL and the Ne Plus Ultra.

24. California State Board of Horticulture, *Biennial Report of the State Board of Horticulture*.

25. California State Board of Horticulture, *Annual Report*, 86.

26. California State Board of Horticulture, *Biennial Report of the State Board of Horticulture*, 326.

27. Wickson, *California Illustrated*, 87.

28. *Sacramento Daily Record-Union*, "Suisun Valley," 8.

29. *Morning Call*, "The Tariff Ghost."

30. California Almond Growers Exchange, *Shall the American Almond Industry Perish?*

31. Goerke-Shrode, "Vanished World of Chinatown"; Leung and Waters, "Chinese Pioneer Families in Suisun Valley."

32. Bancroft, *Chronicles of the Builders of the Commonwealth*.

33. Saxton, *The Indispensable Enemy*.

34. For more on how white supremacy was foundational to California's early statehood, see Almaguer, *Racial Fault Lines*.

35. Gradziel and Martínez-Gómez, "Almond Breeding."

36. US Department of Agriculture, Division of Pomology, *Nut Culture in the United States*, 25.

37. Canfield, "Almond Survey of Butte County," 23.

38. Hasey and Salmon, "Crow Damage to Almonds Increasing."

39. Wood, *Almond Culture in California*, 103:101.

40. Adams and Reed, "Costs of Almond Production in California."

41. Wade, "Biology of the Navel Orangeworm."

42. Wade, 159.

43. Gradziel and Martínez-Gómez, "Shell Seal Breakdown in Almond."

44. Kester, "Almond Cultivar and Breeding Programs in California."

45. Romero, *Economic Poisoning*.

46. Codling, "Environmental Impact and Remediation of Residual Lead and Arsenic Pesticides."

47. Carson, *Silent Spring*.

48. *Encyclopedia of Corporate Social Responsibility*, "Stockholm Convention on Persistent Organic Pollutants (POPs)," edited by Samuel O. Idowu, Nicholas Capaldi, Liangrong Zu, and Ananda Das Gupta, 2013, https://doi.org/10.1007/978-3-642-28036-8_101506.

49. Turusov et al., "Dichlorodiphenyltrichloroethane (DDT)."

50. Naughton and Terry, "Neurotoxicity in Acute and Repeated Organophosphate Exposure."

51. Schnoor, *Fate of Pesticides and Chemicals in the Environment*.

52. Davis, "Toxicity of Organophosphate Chemicals."

53. Goodhue and Klonsky, "Determinants of Adoption of Alternatives to Organophosphate Use."

54. Mueller-Beilschmidt, "Toxicology and Environmental Fate of Synthetic Pyrethroids."

55. Martin-Reina et al., "Insecticide Reproductive Toxicity Profile."

56. Xingmei Liu et al., "Almond Organophosphate and Pyrethroid Use."

57. Van den Bosch, *The Pesticide Conspiracy*.

58. Rominger, "Navel Orangeworm."

59. Schatzki and Ong, "Dependence of Aflatoxin in Almonds," 4513.

60. Lindsey et al., "Methyl Bromide on Dried Fruits and Nuts."

61. Yadav et al., "Acute Aluminum Phosphide Poisoning."

62. Warner, *Agroecology in Action*, 62.

63. US Congress House Committee on Appropriations, "Agriculture-Environmental and Consumer Protection Appropriations," 645.

64. Warner, *Agroecology in Action*, 62.

65. For more on the racial projects underlying agriculture's chemicalization, see Williams and Porter, "Cotton, Whiteness, and Other Poisons."

66. Jill Lindsey Harrison, *Pesticide Drift and the Pursuit of Environmental Justice*.

67. Carravedo Fantova, "El cura de Alquézar."

68. Fusi Aizpurúa and Palafox, *España*.

69. Swyngedouw, "Modernity and Hybridity."

70. Shubert, *Social History of Modern Spain*.

71. Ayerbe Castillo, *Cartilla del cultivo práctico del almendro Desmayo*, 66.

72. Ayerbe Castillo, *El almendro "Desmayo"*, 327. All translations are by the author unless otherwise noted.

73. Ayerbe Castillo, *Cartilla del cultivo práctico del almendro Desmayo*, 9. Original italics.

74. Micke, *Almond Production Manual*.

75. Ayerbe Castillo, *Cartilla del cultivo práctico del almendro Desmayo*, 53.

76. Ayerbe Castillo, 58.

77. Ayerbe Castillo, 66.

78. Carravedo Fantova, "El cura de Alquézar."

79. MacKinnon and Derickson, "From Resilience to Resourcefulness."

80. Sala Roqueta, "Sobre La polinización del almendro Desmayo."

81. Due to Spain's diverse geographic conditions and a decentralized industry, however, no single variety has held a significance on par with California's Nonpareil.

82. All interviews were conducted in confidentiality, and the names of interviewees are withheld by mutual agreement. Transcripts of interviews conducted in Spanish were translated by the author.

83. Almacellas Gort and Marín Sánchez, "Control de plagas y enfermedades en el cultivo del almendro," 69.

84. Barrios i Sanroma and Aymamí, "El futuro de la sanidad vegetal del almendro."

85. Martin et al., "Politics of Care in Technoscience."

86. Grasselly, "Origine et évolution de l'amandier cultivé."

87. Percentage is based on annual production data reported in *United States Department of Agriculture National Agricultural Statistics Service 2016 Certified Organic Survey—California* and by the Almond Board of California.

88. University of California Agriculture and Natural Resources, "Agriculture: Almond Pest Management Guidelines."

2. FLOW: KNOWING PLANT-WATER RELATIONS

1. Almond Board of California, *Almond Almanac 2022*.

2. Doll, "Impacts of Drought on Almond Production."

3. US Climate Data, "Weather History Bakersfield."

4. Stafford, "Almond Survey of Sutter County."

5. While Spain and California have grown distinct varieties of almond, the scale of the increasing water demands cannot be explained by the physiology of the almond tree itself. In California, prominent varieties have been largely unchanged for over a century despite radical shifts in irrigation. In Spain, new varieties gaining popularity often mark a transition from entirely rainfed to modestly irrigated orchards as almonds transition from a complimentary "crop of convenience" to a significant financial investment. While there have been some shifts in the kind of rootstocks (the lower portion of the grafted tree) in use, this shift in genetics is more a result of than a justification for the immense increase in water applications. In fact, the reliance on new forms of nematode-resistant rootstocks can be better understood as a way to cope with the soil-borne pathogens emboldened by rising irrigation. (For more on almond breeding, see chapter 1.)

6. Williams and Murray, "Behind the 'Miracle.'"

7. Siebert et al., "Groundwater Use for Irrigation"; Qiong Hu et al., "Global Cropland Intensification."

8. Guthman, "Back to the Land"; Nickerson et al., "Trends in U.S. Farmland Values."

9. Stoll, *Fruits of Natural Advantage.*

10. Stafford, "Almond Survey of Sutter County."

11. Baird, "Almond Survey of the Acampo Region."

12. University of California, *Report of the College of Agriculture and the Agricultural Experiment Station*, 210.

13. Stafford, "Almond Survey of Sutter County," 43–35.

14. Taylor and Philp, *The Almond in California*, 10.

15. Stoll, *Fruits of Natural Advantage.*

16. Tucker, "Future of the California Almond."

17. Reisman, "Superfood as Spatial Fix."

18. Taylor, *The Almond in California*, 4.

19. California growers today remove orchards after twenty to twenty-five years, but in the early twentieth century trees were expected to last up to fifty years.

20. US Department of Agriculture, National Agricultural Statistics Service, *California Historic Commodity Data.*

21. Garone, *Managing the Garden.*

22. Pisani, *Family Farm to Agribusiness*; Walker, *The Conquest of Bread.*

23. Holmberg and Werenfels, "Water Use of Dry-Farmed Almonds."

24. Taylor and Zilberman, "Diffusion of Drip Irrigation."

25. Venot, "From Obscurity to Prominence."

26. Ghidiu et al., "Drip Chemigation of Insecticides"; Venot, "From Obscurity to Prominence."

27. Sears et al., "Jevons' Paradox and Efficient Irrigation Technology."

28. The 15 percent of growers in this region reportedly using sprinklers (what would now be considered a low-efficiency drag hose system) applied slightly less, 33.6 inches.

29. Taylor and Zilberman, "Diffusion of Drip Irrigation."

30. Phene, "Drip Irrigation Can Reduce California's Water Application."

31. Clemmens et al., "Technical Concepts Related to Conservation of Irrigation."

32. Caswell and Zilberman, "Choices of Irrigation Technologies in California."

33. This widely promoted statistic is likely an overestimate as it represents the portion of respondents participating in a voluntary grower self-assessment called the California Almond Stewardship Program, led by the Almond Board of California. It is more than double the state-wide adoption rate across all farms reported by the California Water Resources Board (California Water Plan Update 2013), a government agency.

34. Doorenbos and Pruitt, *Crop Water Requirements.*

35. Doorenbos and Pruitt, 1.

36. Doorenbos and Pruitt, summary.

37. Herfeld, "Model Transfer in Science."

38. Hanson et al., "Drip Irrigation Provides the Salinity Control."

39. California Department of Water Resources, California Irrigation Management Information System, "Siting Information."

40. This level of landscape standardization is particularly crucial because ET_0 is no longer physically measured by weighing real-time changes in soil moisture on expensive and

cumbersome lysimeter equipment. Instead it is approximated by use of the internationally standardized Penman-Monteith equation.

41. K_c as described here refers to maximum K_c. It reflects four primary factors: crop height, albedo (reflectivity) of crop and soil, canopy resistance to vapor transfer, and evaporation. Because these change over the lifecycle of the plant, K_c is typically represented for each week or month of the plant's life or for a given percentage of ground cover. Because irrigated almonds grow slowly, reaching maturity within four to five years and continuing commercial production through twenty to twenty-five years of age, the maximum summer K_c for mature trees is most relevant to understanding overall water consumption over its lifetime.

42. Guerra et al., "Crop Coefficients."

43. Sanden et al., "California's Effort to Improve Almond Orchard Crop Coefficients."

44. Yufang Jin et al., "Advancing Agricultural Production."

45. Lampinen et al., "Mobile Platform for Measuring Canopy."

46. Johnson et al., "Crop Coefficients for Mature Peach Trees."

47. Krueger et al., "Alternate Year Pruning of Mature 'Nonpareil' Almonds."

48. Rosenstock et al., "Nitrogen Fertilizer Use in California."

49. Muhammad et al., "Optimization of Nitrogen and Potassium Nutrition."

50. Some agronomists argue that high fertilization rates, accompanied by high yields, do not lead to additional contamination. Studies from the same institution concerning rate of nitrogen loss to the soil as percentage of overall application, however, contradict this claim.

51. Goldhamer and Viveros, "Effects of Preharvest Irrigation Cutoff Durations."

52. Sanden, "Fall Irrigation Management."

53. Doll and Shackel, *Drought Tip*.

54. Shackel, "Precision Irrigation Management."

55. Goldhamer and Fereres, "Establishing an Almond Water Production Function."

56. Khaled and Culumber, *Variable Rate Irrigation Practices on Almond*.

57. Under prior flood irrigation, these salts would be periodically flushed down below the root zone by deep saturation, though the salts do not go away; they merely linger lower in the soil profile.

58. When farms build infrastructure to drain salts away from farm fields, the results can be ecologically catastrophic, as was the case for Kesterson Reservoir. See Garone, "Tragedy at Kesterson Reservoir."

59. Almond Board of California, *Almond Salinity Hazard and Leaching Requirements*.

60. Micke, *Almond Production Manual*.

61. One recent analysis shows climate change alone increasing water demand by 4–18 percent across the Mediterranean, particularly for tree crops. Fader et al., "Mediterranean Irrigation Under Climate Change."

62. Serrano et al., "Europe's Orchard."

63. Faulkner, "Gully Erosion."

64. Reisman, "Protecting Provenance, Abandoning Agriculture?"

65. Ministerio de Agricultura, Pesca y Alimentación, *Análisis de las plantaciones de fruto secos*.

66. Ministerio de Agricultura, Pesca y Alimentación, "Superficies y producciones de cultivos."

67. Baker, "Tragedy Exposes Spain's Vast Network of Illegal Wells."

68. Custodio et al., "Sustainability of Intensive Groundwater Development"; Foster et al., "Waterwells."

69. Fornés et al., "Legal Aspects of Groundwater Ownership in Spain."

70. Novo et al., "More Cash and Jobs per Illegal Drop?"

71. Iglesias and Torrents, "Developing High-Density Training Systems."

72. Ibarra, "Drought Wreaks Havoc on Almond Growers."

73. According to USDA Objective Measurements, almond yields dropped by 4.1–7.6 percent during the 2014–15 drought, relative to the average five years prior and jumped back to meet the average in 2016.

74. This is a concern not only for smaller farmers or marginalized groups but for all farmers. Urban water users have a far higher willingness to pay for water, and many agricultural areas are gradually losing ground to urban and exurban development.

75. Fornés et al., "Legal Aspects of Groundwater Ownership in Spain."

76. Molle and Closas, "Why Is State-Centered Groundwater Governance Largely Ineffective?"

77. Dillis et al., "Watering the Emerald Triangle."

78. Méndez-Barrientos et al., "Farmer Participation and Institutional Capture." See also Underhill et al., "Against Settler Sustainability" on the settler-colonial underpinnings of California's groundwater legislation.

3. SYMBIOSIS: PRODUCING POLLINATOR DEPENDENCE

1. Wagner et al., "Insect Decline in the Anthropocene."

2. Ramsey et al., "*Varroa destructor* Feeds Primarily on Honey Bee Fat."

3. Smith et al., "Pathogens, Pests, and Economics."

4. Durant and Otto, "Feeling the Sting?"

5. Goulson et al., "Bee Declines Driven by Combined Stress."

6. Aizen et al., "Global Agricultural Productivity Is Threatened."

7. Food and Agriculture Organization of the United Nations, "Why Bees Matter."

8. Klein et al., "Importance of Pollinators."

9. US Department of Agriculture, "Importance of Pollinators."

10. Aizen et al., "How Much Does Agriculture Depend on Pollinators?"

11. Ghazoul, "Buzziness as Usual?"

12. *Symbiosis* is an umbrella term describing the intimate association between two species, but it need not be beneficial to either. A relationship of mutual benefit would be *mutualism*, one of neutral impact to one party would be *commensalism*, and one of benefit at the others' expense would be *parasitism*. As ecologists well know, these are not inherent or static qualities. A relationship can transition between these states based on the conditions at hand. Mycorrhizal fungi, for example, often act mutualistically, exchanging nutrients with tree roots during times of scarcity, but can also become parasitic when nutrients become abundant. Industrial symbiosis shares this ambiguity. While pollinators and flowering plants are often imagined as mutualists, the notion that honeybees are benefiting from today's almond migration circuit is up for debate.

13. Rucker and Thurman, "Combing the Landscape."

14. Vansell and DeOng, *Survey of Beekeeping in California*, 1.

15. Vansell and DeOng, 2

16. Harvey, "The Spatial Fix."

17. Zierer, "Migratory Beekeepers of Southern California."

18. Cheung, "Fable of the Bees."

19. Canfield, "Almond Survey of Butte County."

20. Taylor and Philp, *The Almond in California*.

21. Tufts and Philp, *Almond Pollination*, 22.

22. Adams, *Cost of Producing Almonds*.

23. Traynor, "History of Almond Pollination."

24. Zierer, "Migratory Beekeepers of Southern California."

25. US Department of Agriculture, Agricultural Marketing Service, *Extracted Honey Grading Manual*.

26. Pellett, *American Honey Plants*.

27. US Department of Agriculture, National Agricultural Statistics Service, *California Historic Commodity Data*.

28. Martin and McGregor, "Changing Trends in Insect Pollination."

29. Kester and Griggs, "Fruit Setting in the Almond."

30. US Department of Agriculture, Agricultural Research Service, "Beekeeping in the United States."

31. Griggs and Iwakiri, "Timing Is Critical for Effective Cross Pollination."

32. "Sample Cost to Produce Almonds: Glenn County 1963."

33. Nolan, "Selection of Honeybee Stock," 314–16.

34. Rucker and Thurman, *Combing the Landscape*.

35. Siebert, "Beekeeping, Pollination, and Externalities."

36. Martin and McGregor, "Changing Trends in Insect Pollination."

37. The USDA marks this transition in its 2021 report; however, this reflects a national average. Nearly all the almond-pollination beekeepers surveyed described the transition happening around 2012–17 when almond pollination saw a rise in fees.

38. Bond et al., "Honey Bees on the Move."

39. Ferrier et al., *Economic Effects and Responses to Changes in Honey Bee Health*.

40. García, "Current Situation on the International Honey Market."

41. Many beekeepers also suggested to me that these USDA numbers were a low estimate based on self-reporting.

42. Cilia, *"It's Never Been Such a Good Time to Be a Beekeeper!"*

43. Martínez-López et al., "Migratory Beekeeping."

44. Commercial hives operate using a Langstroth design, which is a wooden box containing movable frames that facilitate beekeeping procedures and honey removal.

45. Rinkevich, "Detection of Amitraz Resistance."

46. Bond et al., *Honey Bees on the Move*. Honey production per hive dropped 17 percent while the number of hives rose by 7 percent, when comparing the averages of the first and second decades of the twenty-first century.

47. Segrelles Serrano, "La apicultura valenciana."

48. Higes et al., "Assessing the Resistance to Acaricides."

49. López i Gelats, Vallejo Rojas, and Rivera Ferre, "Impactos, vulnerabilidad y adaptación."

50. European Commission, "Honey Market Presentation."

51. Spain's land area is 194,897 square miles (504,782 square kilometers), compared with 2,959,065 square miles (7,663,942 square kilometers) for the contiguous United States.

52. López i Gelats et al., "Impactos, vulnerabilidad y adaptación."

53. Reisman, "Protecting Provenance, Abandoning Agriculture?"

54. Gradziel, "Transfer of Self-Fruitfulness to Cultivated Almond."

55. Socias i Company and Felipe, "Self-Compatibility in Almond."

56. Researchers using genetic analysis (and from a competing research center) later published a study claiming that Guara is in fact the Tuono variety. Dicenta et al., "Origin of the Self-Compatible Almond 'Guara.'"

57. Tierras Apícolas, https://tierrasapicolas.com. As of May 5, 2025, there were only four postings related to facilitating pollination services. Listed on the site are 167 beekeepers, primarily selling their equipment or honey.

58. Champetier et al., "Are the Almond and Beekeeping Industries Gaining Independence?"

59. Champetier et al.

60. Sáez et al., "Bees Increase Crop Yield."

61. According to the USDA 2024 statistics, total bearing almond acreage was estimated to be 1.4 million, and only 2.6 million commercial honeybee hives were counted, meaning two hundred thousand hives shy of the industry optimum.

4. SPACE: CREEPING TOWARD PRECARITY

1. Fitzgerald, *Every Farm a Factory.*

2. Geisseler and Horwath, *Almond Production in California.*

3. Almond Board of California, *Almond Almanac 2022.*

4. Geisseler and Horwath, "Almond Production in California," 2.

5. Baron et al., "Meeting Ecological and Societal Needs for Freshwater."

6. Mann and Dickinson, "Obstacles to the Development of a Capitalist Agriculture."

7. Underhill, "The Return of Pa'ashi."

8. Dillon, "Civilizing Swamps in California"; Nash, *Inescapable Ecologies.*

9. Gates, "Public Land Disposal in California."

10. Ricardo, *Principles of Political Economy and Taxation.*

11. Similar forms of iatrogenic precarity are discussed in Guthman, *Wilted*; Blanchette, *Porkopolis*; Lien, *Becoming Salmon.*

12. Navigant Consulting, "Energy Efficiency Potential and Goals Study."

13. Méndez-Barrientos et al., "Farmer Participation and Institutional Capture."

14. Molle and Closas, "Why Is State-Centered Groundwater Governance Largely Ineffective?"

15. Taxin, "Crop-Rich California Region."

16. Yu Zhan and Minghua Zhang, "Spatial and Temporal Patterns of Pesticide Use."

17. Krueger et al., "Alternate Year Pruning of Mature 'Nonpareil.'"

18. Calla et al., "Selective Sweeps in a Nutshell."

19. Acretrader, a farmland investment platform, quoted California's average irrigated farmland price at $18,600 per acre for 2023. https://acretrader.com/resources/california-farmland-prices.

20. Méndez-Barrientos et al., "Race, Citizenship, and Belonging."

21. London, "Disadvantaged Unincorporated Communities."

22. Egge and Ajibade, "A Community of Fear."

23. Jasechko and Perrone, "California's Central Valley Groundwater Wells Run Dry."

24. Perrone and Jasechko, "Dry Groundwater Wells in the Western United States."

25. Harter et al., "Addressing Nitrate in California's Drinking Water."

26. Balazs et al., "Social Disparities in Nitrate-Contaminated Drinking Water."

27. Levy et al., "Critical Aquifer Overdraft Accelerates Degradation of Groundwater."

28. Smith et al., "Overpumping Leads to California Groundwater Arsenic Threat"; Aiken et al., "Disparities in Drinking Water Manganese Concentrations."

29. Sapbamrer and Hongsibsong, "Effects of Prenatal and Postnatal Exposure to Organophosphate Pesticides."

30. Libre, "Stranded Pesticides." Public health advocates have highlighted how the focus on consumer, rather than occupational, exposure in the national ban callously disregards farmworker safety.

31. Yuzhou Luo and Minghua Zhang, "Spatially Distributed Pesticide Exposure Assessment."

32. Mueller-Beilschmidt, "Toxicology and Environmental Fate of Synthetic Pyrethroids."

33. Wangxin Tang et al., "Pyrethroid Pesticide Residues in the Global Environment."

34. Pellow, "Critical Environmental Justice Studies."

35. Ricart et al., "Modeling the Stakeholder Profile."

36. García-Ruiz, "Effects of Land Uses on Soil Erosion in Spain."

37. Prices cited here are taken from the Lonja de Murcia, one of the three major almond exchanges in Spain establishing daily prices.

38. Reisman, "Superfood as Spatial Fix."

39. The author's analysis is based on data presented in the annual Encuesta sobre superficies y rendimientos de cultivos (Survey on the area and yields of crops) conducted by the Ministerio de Agricultura, Pesca y Alimentación.

40. For grain farmers, the switch to permanent crops meant a much higher upfront investment and also much less flexibility. Unlike annual crops, which can be seeded with something new the next season, almonds are a twenty-five-year commitment. In what economists call "demand hardening," water withdrawals become less responsive to prices because growers know that any reduced irrigation can limit yields for years in the future.

41. World Wildlife Fund, *Illegal Water Use in Spain.*

42. Baker, "Tragedy Exposes Spain's Vast Network of Illegal Wells."

43. Novo et al., "More Cash and Jobs per Illegal Drop?"

44. Sanz et al., "Modeling Aquifer–River Interactions"; Ayuda et al., "Blue Water Footprint."

45. Camacho et al., "Groundwater Extraction Poses Extreme Threat."

46. GeoPortal, "Contenido de nitratos de origen agrario en las aguas subterráneas."

47. Ruiz, "Los nitratos y las aguas subterráneas en España."

48. Doll et al., "Almond Production in Portugal."

49. Olivié and Pérez, "Difficult Escape from Dualism."

50. AgroCLM, "Si no hay almendra." Prices reflect those quoted by the Lonja de Murcia.

51. Reisman, "Protecting Provenance, Abandoning Agriculture?"

52. AgroCLM, "UPA indica que 'el superintensivo machaca al almendro tradicional.'"

53. Ministerio de Medio Ambiente y Medio Rural y Marino, "Encuesta sobre superficies y rendimientos de cultivos."

54. Molino, *La España vacía.*

55. Molinero Hernando, "El espacio rural de España."

56. Pinilla and Sáez, *La despoblación rural en España.*

57. Rivas Sanz and Fernández-Maroto, "Planning for Growth."

58. Cavestany y de Anduaga, "Menos agricultores y mejor agricultura."

59. Van Leeuwen et al., "Evolution of Soil Conservation Policies."

60. Buendía, "Spanish Economic 'Miracle' That Never Was," 51–72.

61. A political platform called Teruel Existe formed in 1999 to demand investments in infrastructure for the neglected rural province and has since gained momentum, staging a protest across twenty-three provinces in 2019 and formalizing itself as a successful political party in 2021.

62. The term *reto demográfico,* though originally conceived of as depopulation, was expanded to include aging and fluctuating populations in order to build a political coalition with coastal regions.

63. Hernández, "Pastoreo contra incendios."

64. Duffy, "War, by Conservation."

65. Sterk and Stoorvogel, "Desertification."

66. La Moncloa, "El Plan de Medidas."

67. Martínez-Abraín et al., "Ecological Consequences of Human Depopulation."

68. García-Ruiz et al., "Rewilding and Restoring Cultural Landscapes."

69. Bosch et al., "Distribution, Abundance and Density of the Wild Boar."

70. European Commission, "Informe sobre el Plan estratégico de la PAC de España."

71. Sáez Pérez, "Análisis de la estrategia nacional frente a la despoblación en el reto demográfico en España," 19.

72. Ramos et al., "Cover Crops Under Different Managements vs. Frequent Tillage"; Cárceles Rodríguez et al., "Soil Management Strategies in Organic Almond Orchards."

73. The name is a portmanteau of the three participating districts, which share ecological and cultural ties; Altiplano de Granada, Los Vélez, and Alto Almanzora.

74. For more on the struggle for dignity in rural Spain, see Franquesa, *Power Struggles.*

75. Brady et al., "Survival and Growth of Organic Farms."

76. Sánchez Martín et al., "Implantación de alojamientos."

77. Molinero Hernando, "El espacio rural de España."

78. Fletcher and Neves, "Contradictions in Tourism."

79. De Groot et al., "Framework for Integrated Ecosystem Services Assessment"; de Leijster et al., "Almond Farm Profitability."

80. Instituto Nacional de Estadística, "Anuario estadistico 2020: Agricultura y ganadería ecológica"; US Department of Agriculture, National Agricultural Statistics Service, *Certified Organic Survey 2021 Summary.*

81. Scales, "Green Capitalism."

5. CONJUNCTURE: ROOTING AGRICULTURAL KNOWLEDGES IN PLACE

1. Robinson, *Black Marxism*; O'Connor, "Two Contradictions of Capitalism."

2. Casert, "EU Scraps Pesticide Proposal."

3. Guthman, "Neoliberalism and the Making of Food Politics."

4. Graddy-Lovelace and Diamond, "From Supply Management to Agricultural Subsidies."

5. Weis, *The Global Food Economy*; Clapp, *Food.*

6. Pulido, "Geographies of Race and Ethnicity."

BIBLIOGRAPHY

Adams, R. L. *Cost of Producing Almonds in California: A Progress Report.* Bulletin 422. College of Agriculture, University of California, 1927. https://play.google.com/books /reader?id=Q1zBwIfVtZgC&pg=GBS.PA2&hl=en.

Adams, R. L., and A.D. Reed. "Costs of Almond Production in California." *California Agriculture,* February 1948. Circular by the Agricultural Experiment Station, College of Agriculture, University of California.

Adams, Vincanne. *Glyphosate and the Swirl: An Agroindustrial Chemical on the Move.* Critical Global Health. Duke University Press, 2023.

Agricultural Economic Insights. "U.S. Almond Production and Consumption Trends." May 17, 2021. https://aei.ag/2021/05/17/united-states-almond-production-consumption-trends.

AgroCLM. "'Si no hay almendra, ¿cómo es que los precios están tan bajos?'" September 13, 2022. https://www.agroclm.com/2022/09/13/si-no-hay-almendra-como-es-que-los -precios-estan-tan-bajos.

AgroCLM. "UPA indica que 'el superintensivo machaca al almendro tradicional.'" June 23, 2021. https://www.agroclm.com/2021/06/23/upa-indica-que-el-superintensivo-machaca -al-almendro-tradicional.

Aiken, Miranda L., Clare E. Pace, Maithili Ramachandran, et al. "Disparities in Drinking Water Manganese Concentrations in Domestic Wells and Community Water Systems in the Central Valley, CA, USA." *Environmental Science & Technology* 57, no. 5 (2023): 1987–96. https://doi.org/10.1021/acs.est.2c08548.

Aizen, Marcelo A., Sebastián Aguiar, Jacobus C. Biesmeijer, et al. "Global Agricultural Productivity Is Threatened by Increasing Pollinator Dependence Without a Parallel Increase in Crop Diversification." *Global Change Biology* 25, no. 10 (2019): 3516–27. https://doi .org/10.1111/gcb.14736.

Aizen, Marcelo A., Lucas A. Garibaldi, Saul A. Cunningham, and Alexandra M. Klein. "How Much Does Agriculture Depend on Pollinators? Lessons from Long-Term Trends in Crop Production." *Annals of Botany* 103, no. 9 (2009): 1579–88. https://doi.org/10.1093/aob/mcp076.

Allen, Gray. *The Almond People.* Blue Diamond Growers, 2000.

Allen, Patricia. *Together at the Table: Sustainability and Sustenance in the American Agrifood System.* Pennsylvania State University Press and Rural Sociological Society, 2004.

Almacellas Gort, Jaume, and Juan Pedro Marín Sánchez. "Control de plagas y enfermedades en el cultivo del almendro." *Vida rural* 332 (September 2011).

Almaguer, Tomás. *Racial Fault Lines: The Historical Origins of White Supremacy in California.* University of California Press, 2009.

Almond Board of California. *Almond Almanac 2022.* Annual report. 2022.

Almond Board of California. *Almond Salinity Hazard and Leaching Requirements.* Document IR20160189. 2016.

Almond Board of California. *Facts About Water Use Among Almond Growers.* Factsheet. May 2014.

Arboleda, Martín. "Revitalizing Science and Technology Studies: A Marxian Critique of More-than-Human Geographies." *Environment and Planning D: Society and Space* 35, no. 2 (2017): 360–78. https://doi.org/10.1177/0263775816664099.

Associated Press. "How China's Taste for Almonds Is Sucking Drought-Stricken California Dry." *South China Morning Post,* April 20, 2015. https://www.scmp.com/news/world/article/1771871/how-chinas-taste-almonds-sucking-drought-stricken-california-dry.

Ayerbe Castillo, Rafael. *Cartilla del cultivo práctico del almendro Desmayo.* Huesca: V. Campo, 1923.

Ayerbe Castillo, Rafael. *El almendro "Desmayo": Su cultivo, terreno, multiplicación, injerto, poda y recolección. Ventajas del "Desmayo" sobre las otras variedades.* Huesca: La Viuda de Justo Martínez, 1922.

Ayuda, María-Isabel, Encarna Esteban, Miguel Martín-Retortillo, and Vicente Pinilla. "The Blue Water Footprint of the Spanish Wine Industry: 1935–2015." *Water* 12, no. 7 (2020): 1872. https://doi.org/10.3390/w12071872.

Baird, William Perry. "An Almond Survey of the Acampo Region, San Joaquin County, California." Bachelor's thesis, College of Agriculture, University of California, 1916. https://hdl.handle.net/2027/uc1.x85917.

Baker, Sam. "Tragedy Exposes Spain's Vast Network of Illegal Wells." *Deutsche Welle,* February 1, 2019. https://www.dw.com/en/spains-vast-network-of-illegal-wells-exposed-after-death-of-toddler/a-47311150.

Balazs, Carolina, Rachel Morello-Frosch, Alan Hubbard, and Isha Ray. "Social Disparities in Nitrate-Contaminated Drinking Water in California's San Joaquin Valley." *Environmental Health Perspectives* 119, no. 9 (2011): 1272–78. https://doi.org/10.1289/ehp.1002878.

Balfour, Sebastian. "Riot, Regeneration and Reaction: Spain in the Aftermath of the 1898 Disaster." *Historical Journal* 38, no. 2 (1995): 405–23. https://doi.org/10.1017/S0018246X00019476.

Bancroft, Hubert Howe. *Chronicles of the Builders of the Commonwealth: Historical Character Study.* History Company, 1892.

Barad, Karen. *Meeting the Universe Halfway: Quantum Physics and the Entanglement of Matter and Meaning.* Duke University Press, 2007.

Barad, Karen. "Posthumanist Performativity: Toward an Understanding of How Matter Comes to Matter." *Signs: Journal of Women in Culture and Society* 28, no. 3 (March 2003): 801–31. https://doi.org/10.1086/345321.

Baron, Jill S., N. LeRoy Poff, Paul L. Angermeier, et al. "Meeting Ecological and Societal Needs for Freshwater." *Ecological Applications* 12, no. 5 (2002): 1247–60. https://doi .org/10.1890/1051-0761(2002)012[1247:MEASNF]2.0.CO;2.

Barrios i Sanroma, G., and A. Aymamí. "El futuro de la sanidad vegetal del almendro." *Revista de fruticultura* 49 (2016): 128–51.

Beckert, Sven. *Empire of Cotton: A Global History.* Vintage Books, 2014.

Benito, Cristóbal Gómez, and Emilio Luque Pulgar. "Modernización agraria, modernización administrativa y franquismo el modelo educativo y administrativo del Servicio de Extensión Agraria (1955–1986)." *Areas: Revista internacional de ciencias sociales* 26 (2007): 131–49.

Blanchette, Alex. *Porkopolis: American Animality, Standardized Life, and the Factory Farm.* Duke University Press, 2020.

Blaut, James M. *The Colonizer's Model of the World: Geographical Diffusionism and Eurocentric History.* Guilford Press, 1993.

Bond, Jennifer K., Claudia Hitaj, David Smith, Kevin Hunt, Agnes Perez, and Gustavo Ferreira. *Honey Bees on the Move: From Pollination to Honey Production and Back.* Economic Research Report 290. Economic Research Service, US Department of Agriculture, June 2021.

Bosch, Jaime, Salvador Peris, Carlos Fonseca, et al. "Distribution, Abundance and Density of the Wild Boar on the Iberian Peninsula, Based on the CORINE Program and Hunting Statistics." *Folia Zoologica* 61, no. 2 (2012): 138–51.

Brady, Michael, David Granatstein, and Elizabeth Kirby. "Survival and Growth of Organic Farms Over the Long-Run." *Journal of the Agricultural and Applied Economics Association* 2 (2023): 248–62.

Brockington, Dan. "Ecosystem Services and Fictitious Commodities." *Environmental Conservation* 38, no. 04 (2011): 367–69. https://doi.org/10.1017/S0376892911000531.

Bronson, Kelly. *The Immaculate Conception of Data: Agribusiness, Activists, and Their Shared Politics of the Future.* McGill–Queen's University Press, 2022.

Brown, Trent. "When Food Regimes Become Hegemonic: Agrarian India through a Gramscian Lens." *Journal of Agrarian Change* 20, no. 1 (2020): 188–206. https://doi.org/10.1111/joac.12344.

Buendia, Luis. "The Spanish Economic 'Miracle' That Never Was." In *Crisis in the Eurozone Periphery*, edited by Owen Parker and Dimitris Tsarouhas. Springer, 2018. https://doi .org/10.1007/978-3-319-69721-5_3.

California Almond Growers Exchange. *Shall the American Almond Industry Perish?* California Almond Growers Exchange, 1921.

California Department of Food and Agriculture. *California Agricultural Statistics Review 2022–2023.* https://www.cdfa.ca.gov/Statistics/PDFs/2022-2023_california_agricultural _statistics_review.pdf.

California Department of Water Resources, California Irrigation Management Information System. "Siting Information." https://cimis.water.ca.gov/Stations.aspx?t=2.

California State Board of Horticulture. *Annual Report.* 1891.

California State Board of Horticulture. *Biennial Report of the State Board of Horticulture of the State of California for 1885 and 1886.* State Office, 1887.

Calla, Bernarda, Mark Demkovich, Joel P. Siegel, et al. "Selective Sweeps in a Nutshell: The Genomic Footprint of Rapid Insecticide Resistance Evolution in the Almond Agroecosystem." *Genome Biology and Evolution* 13, no. 1 (2021): evaa234. https://doi.org/10.1093 /gbe/evaa234.

Camacho, Carlos, Juan J. Negro, Johan Elmberg, et al. "Groundwater Extraction Poses Extreme Threat to Doñana World Heritage Site." *Nature Ecology & Evolution* 6, no. 6 (2022): 654–55. https://doi.org/10.1038/s41559-022-01763-6.

Canfield, C. G. "An Almond Survey of Butte County." Division of Pomology, University of California, 1915. http://hdl.handle.net/2027/uc1.x85968.

Cárceles Rodríguez, Belén, Víctor Hugo Durán Zuazo, Juan Francisco Herencia Galán, et al. "Soil Management Strategies in Organic Almond Orchards: Implications for Soil Rehabilitation and Nut Quality." *Agronomy* 13, no. 3 (2023): 749. https://doi.org/10.3390/agronomy13030749.

Carravedo Fantova, Miguel. "El cura de Alquézar, Gran Cruz del Mérito Agrícola. 1926." In *Añoranzas de Alquézar*. Ayuntamiento de Alquézar, 2017.

Carson, Rachel. *Silent Spring*. 40th anniversary ed., Mariner Books, Houghton Mifflin, 2002.

Casert, Raf. "EU Scraps Pesticide Proposal in Another Concession to Protesting Farmers." Associated Press, February 6, 2024.

Castree, Noel, Catherine Nash, Neil Badmington, Bruce Braun, Jonathon Murdoch, and Sarah Whatmore. "Mapping Posthumanism: An Exchange." *Environment and Planning A* 36, no. 8 (2004): 1341–63. https://doi.org/10.1068/a37127.

Caswell, Margriet, and David Zilberman. "The Choices of Irrigation Technologies in California." *American Journal of Agricultural Economics* 67, no. 2 (1985): 224–34. https://doi.org/10.2307/1240673.

Cavestany y de Anduaga, Rafael. "Menos agricultores y mejor agricultura." *Revista de estudiso agro-sociales*, October 1955, 7–34.

Champetier, Antoine, Hyunok Lee, and Daniel A. Sumner. "Are the Almond and Beekeeping Industries Gaining Independence?" *Choices* 34 (2019): 1–8.

Cheung, Steven N. S. "The Fable of the Bees: An Economic Investigation." *Journal of Law and Economics* 16, no. 1 (1973): 11–33.

Cilia, Laurent. "'It's Never Been Such a Good Time to Be a Beekeeper!' Large-Scale Beekeeping and the Plight of Honey Bees in the United States." PhD thesis, University of Colorado, Boulder, 2020. ProQuest 27958200.

Claire, Theo, and Kevin Surprise. "Moving the Rain: Settler Colonialism, the Capitalist State, and the Hydrologic Rift in California's Central Valley." *Antipode* 54, no. 1 (2022): 153–73. https://doi.org/10.1111/anti.12777.

Clapp, Jennifer. *Food*. 3rd ed. Resources. Polity, 2020.

Clemmens, A. J., R. G. Allen, and C. M. Burt. "Technical Concepts Related to Conservation of Irrigation and Rainwater in Agricultural Systems." *Water Resources Research* 44, no. 7 (2008). https://doi.org/10.1029/2007WR006095.

Cochrane, Willard. *The Development of American Agriculture: A Historical Analysis*. 2nd ed. University of Minnesota Press, 1993.

Codling, Eton. "Environmental Impact and Remediation of Residual Lead and Arsenic Pesticides in Soil." In *Pesticides in the Modern World—Risks and Benefits*, edited by Margarita Stoytcheva. InTech, 2011. https://doi.org/10.5772/17396.

Corburn, Jason. *Street Science: Community Knowledge and Environmental Health Justice*. MIT Press, 2005.

CropLife International. "Advancing Innovation in Agriculture for a Sustainable Future." Accessed 2024. https://croplife.org.

Custodio, Emilio, Andrés Sahuquillo, and José Albiac. "Sustainability of Intensive Ground-water Development: Experience in Spain." *Sustainable Water Resources Management* 5, no. 1 (2019): 11–26. https://doi.org/10.1007/s40899-017-0105-8.

Davis, Frederick Rowe. "The Toxicity of Organophosphate Chemicals." In *Banned: A History of Pesticides and the Science of Toxicology*. Yale University Press, 2014.

De Groot, Rudolf, Simon Moolenaar, Joris de Vente, et al. "Framework for Integrated Ecosystem Services Assessment of the Costs and Benefits of Large Scale Landscape Restoration Illustrated with a Case Study in Mediterranean Spain." *Ecosystem Services* 53 (February 2022): 101383. https://doi.org/10.1016/j.ecoser.2021.101383.

De Leijster, V., R. W. Verburg, M. J. Santos, et al. "Almond Farm Profitability Under Agroecological Management in South-Eastern Spain: Accounting for Externalities and Opportunity Costs." *Agricultural Systems* 183 (August 2020): 102878. https://doi.org/10.1016/j.agsy.2020.102878.

Díaz Geada, Alba, Ana Cabana Iglesia, Lourenzo Fernández Prieto, and Daniel Lanero Táboas. "Agricultural Extension Programmes in Post-War Europe: A Comparative Study of Two Extreme Cases: Spain and the Netherlands (1946–1973)." Paper presented at the 9th European Social Science History Conference, Glasgow, April 11–14, 2012.

Dicenta, F., R. Sánchez-Pérez, M. Rubio, et al. "The Origin of the Self-Compatible Almond 'Guara.'" *Scientia Horticulturae* 197 (December 2015): 1–4. https://doi.org/10.1016/j.scienta.2015.11.005.

Dillis, Christopher, Theodore E. Grantham, Connor McIntee, Bryan McFadin, and Kason Grady. "Watering the Emerald Triangle: Irrigation Sources Used by Cannabis Cultivators in Northern California." *California Agriculture* 73, no. 3 (2019): 146–53. https://doi.org/10.3733/ca.2019a0011.

Dillon, Lindsey. "Civilizing Swamps in California: Formations of Race, Nature, and Property in the Nineteenth Century U.S. West." *Environment and Planning D: Society and Space* 40, no. 2 (2022): 258–75. https://doi.org/10.1177/02637758211026317.

Doll, D. A., P. Serrano, and J. Freire de Andrade. "Almond Production in Portugal: Planting Trends and Production Challenges Within a Developing Industry." *Acta Horticulturae* 1327 (November 2021): 253–60. https://doi.org/10.17660/ActaHortic.2021.1327.33.

Doll, David. "Impacts of Drought on Almond Production." *Growing Produce*, June 10, 2014. https://www.growingproduce.com/fruits/impacts-of-drought-on-almond-production.

Doll, David, and Kenneth Shackel. *Drought Tip: Drought Management for California Almonds*. ANR Publication 8515. Division of Agriculture and Natural Resources, University of California, February 2015. https://doi.org/10.3733/ucanr.8515.

Doorenbos, J., and W. O. Pruitt. *Crop Water Requirements*. Revised ed. FAO Irrigation and Drainage Paper 24. Rome: Food and Agriculture Organization of the United Nations, 1977.

Duffy, Rosaleen. "War, by Conservation." *Geoforum* 69 (February 2016): 238–48. https://doi.org/10.1016/j.geoforum.2015.09.014.

DuPuis, E. Melanie. *Nature's Perfect Food: How Milk Became America's Drink*. New York University Press, 2002.

Durant, Jennie L., and Clint R. V. Otto. "Feeling the Sting? Addressing Land-Use Changes Can Mitigate Bee Declines." *Land Use Policy* 87 (September 2019): 104005. https://doi.org/10.1016/j.landusepol.2019.05.024.

Egge, Michael, and Idowu Ajibade. "A Community of Fear: Emotion and the Hydro-Social Cycle in East Porterville, California." *Journal of Political Ecology* 28, no. 1 (2021). https://doi.org/10.2458/jpe.2366.

European Commission. "Honey Market Presentation: Expert Group 21 April 2022." PowerPoint presentation. https://sotsiaalkindlustusamet.ee/sites/default/files/documents/2022-06/market-presentation-honey_en.pdf.

European Commission. "Informe sobre el Plan estratégico de la PAC de España." 2023–2027. https://www.mapa.gob.es/es/pac/pac-2023-2027/pepac-33_tcm30-704028.pdf.

Eurostat. "Beehive numbers in EU: Agriculture's buzzing backbone." May 2023. https://ec.europa.eu/eurostat/web/products-eurostat-news/w/EDN-20230519-1.

Evans, James, and Phil Jones. "The Walking Interview: Methodology, Mobility and Place." *Applied Geography* 31, no. 2 (2011): 849–58.

Fader, M., S. Shi, W. von Bloh, A. Bondeau, and W. Cramer. "Mediterranean Irrigation Under Climate Change: More Efficient Irrigation Needed to Compensate for Increases in Irrigation Water Requirements." *Hydrology and Earth System Sciences* 20, no. 2 (2016): 953–73. https://doi.org/10.5194/hess-20-953-2016.

Fairbairn, Madeleine. "Framing Transformation: The Counter-Hegemonic Potential of Food Sovereignty in the US Context." *Agriculture and Human Values* 29, no. 2 (2012): 217–30. https://doi.org/10.1007/s10460-011-9334-x.

Faulkner, H. "Gully Erosion Associated with the Expansion of Unterraced Almond Cultivation in the Coastal Sierra de Lujar, S. Spain." *Land Degradation & Rehabilitation* 6, no. 3 (1995): 179–200.

Faunt, Claudia C., Michelle Sneed, Jon Traum, and Justin T. Brandt. "Water Availability and Land Subsidence in the Central Valley, California, USA." *Hydrogeology Journal* 24, no. 3 (2016): 675–84. https://doi.org/10.1007/s10040-015-1339-x.

Federación española de asociaciones de productores exportadores de frutas, hortalizas, flores y plantas vivas. "Exportación/importación españolas de frutas y hortalizas." Accessed 2022. https://www.fepex.es/datos-del-sector/exportacion-importacion-espa%C3%B1ola-frutas-hortalizas.

Ferrier, Peyton M., Randal Rucker, Walter Thurman, and Michael Burgett. *Economic Effects and Responses to Changes in Honey Bee Health.* Economic Research Report 246. Economic Research Service, US Department of Agriculture, March 2018.

Fitzgerald, Deborah. *The Business of Breeding: Hybrid Corn in Illinois, 1890–1940.* Cornell University Press, 1990.

Fitzgerald, Deborah. *Every Farm a Factory: The Industrial Ideal in American Agriculture.* Yale University Press, 2010.

Fletcher, Robert, and Katja Neves. "Contradictions in Tourism: The Promise and Pitfalls of Ecotourism as a Manifold Capitalist Fix." *Environment and Society* 3, no. 1 (2012). https://doi.org/10.3167/ares.2012.030105.

Fornés, Juan María, África De La Hera, Ramón Llamas, and Pedro Martínez-Santos. "Legal Aspects of Groundwater Ownership in Spain." *Water International* 32, no. 4 (2007): 676–84. https://doi.org/10.1080/02508060.2007.9709698.

Food and Agriculture Organization of the United Nations. *Why Bees Matter: The Importance of Bees and Other Pollinators for Food and Agriculture.* FAO, 2018.

Foster, Stephen, Ricardo Hirata, and Emilio Custodio. "Waterwells: How Can We Make Legality More Attractive?" *Hydrogeology Journal* 29, no. 4 (2021): 1365–68. https://doi.org/10.1007/s10040-021-02319-x.

Fox, Nick J., and Pam Alldred. "New Materialist Social Inquiry: Designs, Methods and the Research-Assemblage." *International Journal of Social Research Methodology* 18, no. 4 (2015): 399–414. https://doi.org/10.1080/13645579.2014.921458.

Franquesa, Jaume. *Power Struggles: Dignity, Value, and the Renewable Energy Frontier in Spain*. New Anthropologies of Europe. Indiana University Press, 2018.

Freidberg, Susanne. *French Beans and Food Scares: Culture and Commerce in an Anxious Age*. Oxford University Press, 2004.

Freidberg, Susanne. *Fresh: A Perishable History*. Belknap Press of Harvard University Press, 2009.

Friedland, William H., Amy E. Barton, and Robert J. Thomas. *Manufacturing Green Gold: Capital, Labor, and Technology in the Lettuce Industry*. Arnold and Caroline Rose Monograph Series of the American Sociological Association. Cambridge University Press, 1981.

Fruit Logistica. *European Statistics Handbook 2023*. Germany: Messe Berline GmbH, 2023.

Fusi Aizpurúa, Juan Pablo, and Jordi Palafox. *España: 1808–1996: El desafío de la modernidad*. 5th ed. Madrid: Espasa, 2003.

García Moreno, F. Domingo. *El almendro: Su origen, utilidad, clima, terreno, variedades, multiplicación, cuidados, trasplante, injerto, poda, abono, recolección, conservación y aplicaciones, accidentes y enemigos*. Madrid: Librería de Cuesta, 1912.

García, Norberto L. "The Current Situation on the International Honey Market." *Bee World* 95, no. 3 (2018): 89–94. https://doi.org/10.1080/0005772X.2018.1483814.

García-Ruiz, J. M., T. Lasanta, E. Nadal-Romero, N. Lana-Renault, and B. Álvarez-Farizo. "Rewilding and Restoring Cultural Landscapes in Mediterranean Mountains: Opportunities and Challenges." *Land Use Policy* 99 (December 2020): 104850. https://doi.org/10.1016/j.landusepol.2020.104850.

García-Ruiz, José M. "The Effects of Land Uses on Soil Erosion in Spain: A Review." *Catena* 81, no. 1 (April 2010): 1–11. https://doi.org/10.1016/j.catena.2010.01.001.

Garone, Philip. *Managing the Garden: Agriculture, Reclamation, and Restoration in the Sacramento-San Joaquin Delta*. Delta Narratives Project. California Delta Protection Commission, 2015. https://cawaterlibrary.net/document/delta-narratives-managing-the-garden-agriculture-reclamation-and-restoration-in-the-sacramento-san-joaquin-delta.

Garone, Philip. "The Tragedy at Kesterson Reservoir: A Case Study in Environmental History and a Lesson in Ecological Complexity." *Environs* 22, no. 2 (1999): 107–44.

Gates, Paul W. "Public Land Disposal in California." *Agricultural History, Agriculture in the Development of the Far West: A Symposium* 49, no. 1 (1975): 158–78.

Geisseler, Daniel, and William R. Horwath. Almond Production in California. University of California, Davis, June 2016. https://apps1.cdfa.ca.gov/FertilizerResearch/docs/Almond_Production_CA.pdf.

GeoPortal. "Contenido de nitratos de origen agrario en las aguas subterráneas." Ministerio para la Transición Ecológica y el Reto Demográfico, 2022. https://sig.mapama.gob.es/geoportal.

Ghazoul, J. "Buzziness as Usual? Questioning the Global Pollination Crisis." *Trends in Ecology & Evolution* 20, no. 7 (2005): 367–73. https://doi.org/10.1016/j.tree.2005.04.026.

Ghidiu, Gerald, Thomas Kuhar, John Palumbo, and David Schuster. "Drip Chemigation of Insecticides as a Pest Management Tool in Vegetable Production." *Journal of Integrated Pest Management* 3, no. 3 (2012): E1–5. https://doi.org/10.1603/IPM10022.

Gibson-Graham, J. K. *The End of Capitalism (as We Knew It): A Feminist Critique of Political Economy.* University of Minnesota Press, 2006.

Goerke-Shrode, Sabine. "A Look into Vanished World of Chinatown." *The Reporter,* May 1, 2005.

Goldhamer, David A., and Elias Fereres. "Establishing an Almond Water Production Function for California Using Long-Term Yield Response to Variable Irrigation." *Irrigation Science* 35, no. 3 (2017): 169–79. https://doi.org/10.1007/s00271-016-0528-2.

Goldhamer, David A., and Mario Viveros. "Effects of Preharvest Irrigation Cutoff Durations and Postharvest Water Deprivation on Almond Tree Performance." *Irrigation Science* 19, no. 3 (2000): 125–31. https://doi.org/10.1007/s002710000013.

Goodhue, Rachael E., and Karen Klonsky. "Determinants of Adoption of Alternatives to Organophosphate Use in California Almonds." Paper presented at the annual meeting of the American Agricultural Economics Association, Denver, CO, August 1–4, 2004.

Gorman, Hugh S. *The Story of N: A Social History of the Nitrogen Cycle and the Challenge of Sustainability.* Studies in Modern Science, Technology, and the Environment. Rutgers University Press, 2013.

Goulson, Dave, Elizabeth Nicholls, Cristina Botías, and Ellen L. Rotheray. "Bee Declines Driven by Combined Stress from Parasites, Pesticides, and Lack of Flowers." *Science* 347, no. 6229 (2015): 1255957.

Graddy-Lovelace, Garrett, and Adam Diamond. "From Supply Management to Agricultural Subsidies—and Back Again? The U.S. Farm Bill and Agrarian (In)Viability." *Journal of Rural Studies* 50 (February 2017): 70–83. https://doi.org/10.1016/j.jrurstud.2016.12.007.

Gradziel, Thomas M. "Transfer of Self-Fruitfulness to Cultivated Almond from Peach and Wild Almond." *Horticulturae* 8, no. 10 (2022): 965. https://doi.org/10.3390/horticulturae8100965.

Gradziel, Thomas M., and Pedro Martínez-Gómez. "Almond Breeding." In *Plant Breeding Reviews,* vol. 37, ed. Jules Janick, 207–58. Wiley-Blackwell, 2013.

Gradziel, T. M., and P. Martínez-Gómez. "Shell Seal Breakdown in Almond Is Associated with the Site of Secondary Ovule Abortion." *Journal of the American Society for Horticultural Science* 127, no. 1 (2002): 69–74. https://doi.org/10.21273/JASHS.127.1.69.

Grasselly, C. "Réflexions diverses sur l'évolution des objectifs d'amélioration de l'amandier." *Options méditerranéennes* (1984): 147–50. Paris: CIHEAM, 1984.

Grasselly, C. "Origine et évolution de l'amandier cultivé." *Options méditerranéennes* 32 (1976): 45–49.

Griggs, W. H., and B. T. Iwakiri. "Timing Is Critical for Effective Cross Pollination of Almond Flowers." *California Agriculture,* January 1964: 6–7.

Guerra, E., F. Ventura, and R. L. Snyder. "Crop Coefficients: A Literature Review." *Journal of Irrigation and Drainage Engineering* 142, no. 3 (2016): 06015006. https://doi.org/10.1061/(ASCE)IR.1943-4774.0000983.

Guthman, Julie. "Back to the Land: The Paradox of Organic Food Standards." *Environment and Planning A: Economy and Space* 36, no. 3 (2004): 511–28. https://doi.org/10.1068/a36104.

Guthman, Julie. "Neoliberalism and the Making of Food Politics in California." *Geoforum* 39, no. 3 (2008): 1171–83. https://doi.org/10.1016/j.geoforum.2006.09.002.

Guthman, Julie. *Wilted: Pathogens, Chemicals, and the Fragile Future of the Strawberry Industry.* Critical Environments: Nature, Science, and Politics 6. University of California Press, 2019.

Hamblin, James. "The Dark Side of Almond Use." *The Atlantic*, August 28, 2014. https://www.theatlantic.com/health/archive/2014/08/almonds-demon-nuts/379244.

Hanson, Blaine R., Don E. May, Jirka Simůnek, Jan W. Hopmans, and Robert B. Hutmacher. "Drip Irrigation Provides the Salinity Control Needed for Profitable Irrigation of Tomatoes in the San Joaquin Valley." *California Agriculture* 63, no. 3 (2009): 131–36. https://doi.org/10.3733/ca.v063n03p131.

Haraway, Donna J. *Manifestly Haraway*. Posthumanities 37. University of Minnesota Press, 2016.

Haraway, Donna J. *Modest_Witness@Second_Millennium. FemaleMan_Meets_OncoMouse: Feminism and Technoscience*. Routledge, 1997.

Haraway, Donna. "Situated Knowledges: The Science Question in Feminism and the Privilege of Partial Perspective." *Feminist Studies* 14, no. 3 (1988): 575. https://doi.org/10.2307/3178066.

Haraway, Donna Jeanne, and Thyrza Nichols Goodeve. *How Like a Leaf: An Interview with Thyrza Nichols Goodeve*. Routledge, 2000.

Harrison, Jill Lindsey. *Pesticide Drift and the Pursuit of Environmental Justice*. Food, Health, and the Environment. MIT Press, 2011.

Harrison, Joseph. "The Agrarian History of Spain, 1800–1960." *Agricultural History Review* 37, no. 2 (1989): 180–87.

Hart, Gillian. "Relational Comparison Revisited: Marxist Postcolonial Geographies in Practice." *Progress in Human Geography* 42, no. 3 (2018): 371–94. https://doi.org/10.1177/0309132516681388.

Harter, Thomas, Jay R. Lund, Jeannie Darby, et al. "Addressing Nitrate in California's Drinking Water: Report for the State Water Resources Control Board Report to the Legislature." Groundwater Nitrate Project, Implementation of Senate Bill X2 1. Center for Watershed Sciences, University of California, Davis, 2012. http://groundwaternitrate.ucdavis.edu.

Harvey, David. "The Spatial Fix—Hegel, Von Thunen, and Marx." *Antipode* 13, no. 3 (1981): 1–12.

Hasey, Janine, and Terrell P. Salmon. "Crow Damage to Almonds Increasing; No Foolproof Solution in Sight." *California Agriculture* 47, no. 5 (1993): 21–23.

Henke, Christopher. *Cultivating Science, Harvesting Power: Science and Industrial Agriculture in California*. Inside Technology. MIT Press, 2008.

Herfeld, Catherine. "Model Transfer in Science." In *The Routledge Handbook of Philosophy of Scientific Modeling*, edited by Tarja Knuuttila, Natalia Carrillo, and Rami Koskinen. Routledge Handbooks in Philosophy. Routledge, Taylor & Francis, 2025.

Hernández, Lourdes. "Pastoreo contra incendios: Propuesta de WWF España para adaptar el territorio al cambio climático." WWF España, 2022.

Hetherington, Kregg. *The Government of Beans: Regulating Life in the Age of Monocrops*. Duke University Press, 2020.

Heynen, Nik, and Paul Robbins. "The Neoliberalization of Nature: Governance, Privatization, Enclosure and Valuation." *Capitalism Nature Socialism* 16, no. 1 (2005): 5–8. https://doi.org/10.1080/1045575052000335339.

Higes, Mariano, Raquel Martín-Hernández, Carmen Sara Hernández-Rodríguez, and Joel González-Cabrera. "Assessing the Resistance to Acaricides in *Varroa destructor* from Several Spanish Locations." *Parasitology Research* 119, no. 11 (2020): 3595–601. https://doi.org/10.1007/s00436-020-06879-x.

Holmberg, David, and Lukas Werenfels. "Water Use of Dry-Farmed Almonds Under Clean and Noncultivation Conditions." *California Agriculture* 21, no. 9 (1967): 6–7.

Holt-Giménez, Eric. *Campesino a Campesino: Voices from Latin America's Farmer to Farmer Movement for Sustainable Agriculture.* Food First Books, 2006.

Ibarra, Rich. "Drought Wreaks Havoc on Almond Growers as Water Prices Skyrocket." Capital Public Radio, May 7, 2014. http://www.capradio.org/articles/2014/05/07/drought -wreaks-havoc-on-almond-growers-as-water-prices-skyrocket.

Iglesias, I., and J. Torrents. "Developing High-Density Training Systems in Prunus Tree Species for an Efficient and Sustainable Production." *Acta Horticulturae* 1346 (September 2022): 219–28. https://doi.org/10.17660/ActaHortic.2022.1346.28.

Jackson, Robert H., and Edward D. Castillo. *Indians, Franciscans, and Spanish Colonization: The Impact of the Mission System on California Indians.* University of New Mexico Press, 1996.

Jasechko, Scott, and Debra Perrone. "California's Central Valley Groundwater Wells Run Dry During Recent Drought." *Earth's Future* 8, no. 4 (2020). https://doi.org/10.1029 /2019EF001339.

Jasechko, Scott, Hansjörg Seybold, Debra Perrone, Ying Fan, Mohammad Shamsudduha, Richard G. Taylor, Othman Fallatah, and James W. Kirchner. "Rapid Groundwater Decline and Some Cases of Recovery in Aquifers Globally." *Nature* 625, no. 7996 (2024): 715–21. https://doi.org/10.1038/s41586-023-06879-8.

Johnson, R. S., J. Ayars, T. Trout, R. Mead, and C. Phene. "Crop Coefficients for Mature Peach Trees Are Well Correlated with Midday Canopy Light Interception." *Acta Horticulturae,* no. 537 (October 2000): 455–60. https://doi.org/10.17660/ActaHortic.2000.537.53.

Kester, Dale E. "Almond Cultivar and Breeding Programs in California." *Acta Horticulturae* 373 (September 1994): 13–28. https://doi.org/10.17660/ActaHortic.1994.373.1.

Kester, Dale E., and W. H. Griggs. "Fruit Setting in the Almond: The Pattern of Flower and Fruit Drop." In *Proceedings of the American Society for Horticultural Science,* vol. 74, edited by J. R. Magness, 214–19. American Society for Horticultural Science, 1959.

Khaled, Bali, and Catherine Culumber. *Variable Rate Irrigation Practices on Almond.* Annual research report. Almond Board of California, 2018.

Kimura, Aya Hirata. *Radiation Brain Moms and Citizen Scientists: The Gender Politics of Food Contamination after Fukushima.* Duke University Press, 2016.

Kinchy, Abby. *Seeds, Science, and Struggle: The Global Politics of Transgenic Crops.* MIT Press, 2012.

Klein, A.-M., B. E Vaissiere, J. H. Cane, et al. "Importance of Pollinators in Changing Landscapes for World Crops." *Proceedings of the Royal Society B: Biological Sciences* 274, no. 1608 (2007): 303–13. https://doi.org/10.1098/rspb.2006.3721.

Kloppenburg, Jack Ralph, Jr. *First the Seed: The Political Economy of Plant Biotechnology.* 2nd ed. Science and Technology in Society. University of Wisconsin Press, 2004.

Krueger, W. H., Warren C. Micke, and James Yeager. "Alternate Year Pruning of Mature 'Nonpareil' Almonds (*Prunus communis*)." *Acta Horticulturae* 470 (1998): 473–76. https://doi.org/10.17660/ActaHortic.1998.470.66.

La Moncloa. "El Plan de Medidas ante el Reto Demográfico destinará más de 10.000 millones y 130 políticas activas a luchar contra la despoblación y garantizar la cohesión territorial y social." Gobierno de España, March 16, 2021. https://www.lamoncloa.gob .es/serviciosdeprensa/notasprensa/transicion-ecologica/Paginas/2021/160321-plan-reto -demografico.aspx.

Lampinen, Bruce D., Vasu Udompetaikul, Gregory T. Browne, et al. "A Mobile Platform for Measuring Canopy Photosynthetically Active Radiation Interception in Orchard Systems." *HortTechnology* 22, no. 2 (2012): 237–44. https://doi.org/10.21273/HORT TECH.22.2.237.

Latour, Bruno. *The Pasteurization of France*. Translated by John Law and Alan Sheridan. Harvard University Press, 1984.

Law, John. *Material Semiotics*. UK and Norway: The Open University and Sámi Allaskuvla, 2019. www.heterogeneities.net/publications/Law2019MaterialSemiotics.pdf.

Legun, Katharine. "Tiny Trees for Trendy Produce: Dwarfing Technologies as Assemblage Actors in Orchard Economies." *Geoforum* 65 (October 2015): 314–22. https://doi.org /10.1016/j.geoforum.2015.03.009.

Leung, Peter C. Y., and Tony Waters. "Chinese Pioneer Families in Suisun Valley (1870– 1980)." In *150 Years of the Chinese Presence in California: Honor the Past, Engage the Present, Build the Future* = 從金山到千禧的風和雨, 520. Sacramento Chinese Culture Foundation and Department of Asian-American Studies, University of California, Davis, 2001.

Levy, Zeno F., Bryant C. Jurgens, Karen R. Burow, et al. "Critical Aquifer Overdraft Accelerates Degradation of Groundwater Quality in California's Central Valley During Drought." *Geophysical Research Letters* 48, no. 17 (2021). https://doi.org/10.1029/2021GLo 94398.

Libre, Christina. "'Stranded Pesticides': U.S. Agricultural Worker Vulnerability in the Wake of the 2021 Chlorpyrifos Food Ban." *Ecology Law Quarterly* 49, no. 2 (2022). https://doi .org/10.15779/Z38HD7NT6V.

Lien, Marianne Elisabeth. *Becoming Salmon: Aquaculture and the Domestication of a Fish*. California Studies in Food and Culture 55. University of California Press, 2015.

Lindsay, Brendan C. *Murder State: California's Native American Genocide, 1846—1873*. University of Nebraska Press, 2012.

Lindsey, Patricia J., Susie S. Briggs, Adel A. Kader, and Kirby Moulton. "Methyl Bromide on Dried Fruits and Nuts: Issues and Alternatives." In *Chemical Use in Food Processing and Postharvest Handling: Issues and Alternatives*. Agricultural Issues Center, University of California, Davis, 1989.

London, Jonathan K. "Disadvantaged Unincorporated Communities and the Struggle for Water Justice in California" *Water Alternatives* 14, no. 2 (2021).

López i Gelats, Feliu, Virginia Vallejo Rojas, and Marta Guadalupe Rivera Ferre. "Impactos, vulnerabilidad y adaptación al cambio climático de la apicultura mediterránea." Ministerio de Agricultura y Pesca, Alimentación y Medio Ambiente, October 2016.

Lowe, Lisa. "Insufficient Difference." *Ethnicities* 5, no. 3 (2005): 409–14. https://doi.org/10.1177 /14687968050050308.

MacKinnon, Danny, and Kate Driscoll Derickson. "From Resilience to Resourcefulness: A Critique of Resilience Policy and Activism." *Progress in Human Geography* 37, no. 2 (2013): 253–70. https://doi.org/10.1177/0309132512454775.

Madley, Benjamin. *An American Genocide: The United States and the California Indian Catastrophe, 1846-1873*. Lamar Series in Western History. Yale University Press, 2016.

Mann, Susan A., and James M. Dickinson. "Obstacles to the Development of a Capitalist Agriculture." *Journal of Peasant Studies* 5, no. 4 (1978): 466–81. https://doi.org/10.1080 /03066157808438058.

Martin, Aryn, Natasha Myers, and Ana Viseu. "The Politics of Care in Technoscience." *Social Studies of Science* 45, no. 5 (2015): 625–41. https://doi.org/10.1177/0306312715602073.

Martin, E. C., and S. E. McGregor. "Changing Trends in Insect Pollination of Commercial Crops." *Annual Review of Entomology* 18, no. 1 (1973): 207–26. https://doi.org/10.1146/annurev.en.18.010173.001231.

Martínez-Abraín, Alejandro, Juan Jiménez, Ignacio Jiménez, et al. "Ecological Consequences of Human Depopulation of Rural Areas on Wildlife: A Unifying Perspective." *Biological Conservation* 252 (December 2020): 108860. https://doi.org/10.1016/j.biocon.2020.108860.

Martínez-López, Vicente, Carlos Ruiz, and Pilar De la Rúa. "Migratory Beekeeping and Its Influence on the Prevalence and Dispersal of Pathogens to Managed and Wild Bees." *International Journal for Parasitology: Parasites and Wildlife* 18 (August 2022): 184–93. https://doi.org/10.1016/j.ijppaw.2022.05.004.

Martin-Reina, Jose, José A. Duarte, Lucas Cerrilos, Juan D. Bautista, and Isabel Moreno. "Insecticide Reproductive Toxicity Profile: Organophosphate, Carbamate and Pyrethroids." *Journal of Toxins* 4, no. 1 (2017): 7.

Mathews, Andrew S. "Landscapes and Throughscapes in Italian Forest Worlds: Thinking Dramatically about the Anthropocene." *Cultural Anthropology* 33, no. 3 (2018): 386–414. https://doi.org/10.14506/ca33.3.05.

Mathews, Andrew S. *Trees Are Shape Shifters: How Cultivation, Climate Change, and Disaster Create Landscapes.* Yale University Press, 2022.

McMichael, Philip. "Incorporating Comparison Within a World-Historical Perspective: An Alternative Comparative Method." *American Sociological Review* 55, no. 3 (1990): 385–97.

Méndez-Barrientos, Linda E., Amanda L. Fencl, Cassandra L. Workman, and Sameer H. Shah. "Race, Citizenship, and Belonging in the Pursuit of Water and Climate Justice in California." *Environment and Planning E: Nature and Space* 6, no. 3 (2023): 1614–35. https://doi.org/10.1177/25148486221133282.

Méndez-Barrientos, Linda Estelí, Alyssa DeVincentis, et al. "Farmer Participation and Institutional Capture in Common-Pool Resource Governance Reforms. The Case of Groundwater Management in California." *Society & Natural Resources* 33, no. 12 (2020): 1486–1507. https://doi.org/10.1080/08941920.2020.1756548.

Micke, Warren C., ed. *Almond Production Manual.* Division of Agriculture and Natural Resources, University of California, 1996.

Ministerio de Agricultura. *El almendro: Su importancia y cultivo.* Madrid: Ministerio de Agricultura, Dirección General de Agricultura, 1963.

Ministerio de Agricultura, Pesca y Alimentación. "Agricultura Ecológica" In *Anuario de estadística.*" Madrid: Gobierno de España, 2020.

Ministerio de Agricultura, Pesca y Alimentación. *Análisis de las plantaciones de fruto seco 2005: Encuesta sobre superficies y rendimientos de cultivos.* 2005.

Ministerio de Agricultura, Pesca y Alimentación. "Superficies y producciones de cultivos." In *Anuario de estadística 2020.* Ministerio de Agricultura, Pesca y Alimentación. Madrid: 2021.

Ministerio de Medio Ambiente y Medio Rural y Marino, *Encuesta sobre superficies y rendimientos de cultivos: Resultados 2010.* Secretaría General Técnica: 2011.

Mintz, Sidney W. *Sweetness and Power: The Place of Sugar in Modern History.* Penguin, 1986.

Mitchell, Timothy. "Fixing the Economy." *Cultural Studies* 12, no. 1 (1998): 82–101. https://doi.org/10.1080/095023898335627.

Mol, Annemarie. *The Body Multiple: Ontology in Medical Practice*. Science and Cultural Theory. Duke University Press, 2002.

Molinero Hernando, Fernando. "El espacio rural de España: Evolución, delimitación y clasificación." *Cuadernos geográficos* 58, no. 3 (2019): 19–56. https://doi.org/10.30827/cuadgeo.v58i3.8643.

Molino, Sergio del. *La España vacía: Viaje por un país que nunca fue*. Colección Noema. Madrid: Turner, 2016.

Molle, François, and Alvar Closas. "Why Is State-Centered Groundwater Governance Largely Ineffective? A Review." *WIREs Water* 7, no. 1 (2020): e1395. https://doi.org/10.1002/wat2.1395.

Morning Call. "The Tariff Ghost: California Fruit-Growers Alarmed." July 14, 1893.

Mueller-Beilschmidt, Doria. "Toxicology and Environmental Fate of Synthetic Pyrethroids." *Journal of Pesticide Reform* 10, no. 3 (1990).

Muhammad, Saiful, Blake L. Sanden, Sebastian Saa, et al. "Optimization of Nitrogen and Potassium Nutrition to Improve Yield and Yield Parameters of Irrigated Almond (*Prunus dulcis* (Mill.) D. A. Webb)." *Scientia Horticulturae* 228 (January 2018): 204–12. https://doi.org/10.1016/j.scienta.2017.10.024.

Murray, Ivan, Gabriel Jover-Avellà, Onofre Fullana, and Enric Tello. "Biocultural Heritages in Mallorca: Explaining the Resilience of Peasant Landscapes Within a Mediterranean Tourist Hotspot, 1870–2016." *Sustainability* 11, no. 7 (2019): 1926. https://doi.org/10.3390/su11071926.

Nash, Linda Lorraine. *Inescapable Ecologies: A History of Environment, Disease, and Knowledge*. University of California Press, 2006.

Naughton, Sean X., and Alvin V. Terry Jr. "Neurotoxicity in Acute and Repeated Organophosphate Exposure." *Toxicology* 408 (September 2018): 101–12. https://doi.org/10.1016/j.tox.2018.08.011.

Navigant Consulting. "Energy Efficiency Potential and Goals Study for 2015 and Beyond." California Public Utilities Commission, 2015. http://www.cpuc.ca.gov/General.aspx?id=6442452620.

Navon, Daniel. "Reiterated Fact-Making: Explaining Transformation and Continuity in Scientific Facts." *American Sociological Review* 89, no. 5 (2024): 849–75. https://doi.org/10.1177/00031224241271100.

Nickerson, Cynthia, Mitchell Morehart, Todd Kuethe, Jayson Beckman, Jennifer Ifft, and Ryan Williams. "Trends in U.S. Farmland Values and Ownership." Economic Information Bulletin. Economic Research Service, US Department of Agriculture, February 2012. https://digitalcommons.unl.edu/usdaarsfacpub/1598.

Nolan, W. J. "Selection of Honeybee Stock Is Important to Beekeeper and Orchardist." In *Yearbook of Agriculture 1934*, ed. Milton S. Eisenhower and Arthur P. Chew. Bureau of Entomology, US Department of Agriculture, 1934.

Norton, Jack. *Genocide in Northwestern California: When Our Worlds Cried*. Indian Historian Press, 1979.

Novo, Paula, Aurélien Dumont, Bárbara A. Willaarts, and Elena López-Gunn. "More Cash and Jobs per Illegal Drop? The Legal and Illegal Water Footprint of the Western Mancha

Aquifer (Spain)." *Environmental Science & Policy* 51 (August 2015): 256–66. https://doi
.org/10.1016/j.envsci.2015.04.013.

O'Connor, James. "On the Two Contradictions of Capitalism." *Capitalism Nature Socialism* 2,
no. 3 (1991): 107–9. https://doi.org/10.1080/10455759109358463.

Ogden, Laura A., Billy Hall, and Kimiko Tanita. "Animals, Plants, People, and Things:
A Review of Multispecies Ethnography." *Environment and Society* 4, no. 1 (2013). https://
doi.org/10.3167/ares.2013.040102.

Olivié, Iliana, and Aitor Pérez. "The Difficult Escape from Dualism: The Green Morocco
Plan at a Crossroads." *New Medit: A Mediterranean Journal of Economics, Agriculture
and Environment* 17, no. 3 (2018): 37–50. https://doi.org/10.30682/nm1803d.

Palomino, Joaquin. "California's Thirsty Almonds: How the Water-Intensive Crop Is Helping Drive the Governor's $25 Billion Plan to Ship Water to the Desert." *East Bay Express*,
February 5, 2014. https://eastbayexpress.com/californias-thirsty-almonds-1.

Panelli, R. "More-than-Human Social Geographies: Posthuman and Other Possibilities."
Progress in Human Geography 34, no. 1 (February 1, 2010): 79–87. https://doi.org/10.1177
/0309132509105007.

Pellett, Frank C. *American Honey Plants: Together with Those Which Are of Special Value to
the Beekeeper as a Source of Pollen.* 2nd ed. American Bee Journal, 1923.

Pellow, David N. "Critical Environmental Justice Studies." In *Environmental Justice*, edited
by Brendan Coolsaet. Key Issues in Environment and Sustainability. Routledge, 2020.
https://doi.org/10.4324/9780429029585-26.

Perrone, D., and S. Jasechko. "Dry Groundwater Wells in the Western United States." *Environmental Research Letters* 12, no. 10 (2017): 104002. https://doi.org/10.1088/1748-9326
/aa8aco.

Phene, C. J. "Drip Irrigation Can Reduce California's Water Application by 2.4x106 Acre-
Ft per Year Without Yield Reduction." Paper presented at the 5th National Decennial
Irrigation Conference, Phoenix, Arizona, 2010.

Philpott, Tom. "Lay Off the Almond Milk, You Ignorant Hipsters." *Mother Jones*, July 16,
2014.   https://www.motherjones.com/food/2014/07/lay-off-almond-milk-ignorant-hip
sters.

Pinilla, Vicente, and Luis Antonio Sáez. *La despoblación rural en España: Génesis de un
problema y políticas innovadoras.* Áreas Escasamente Pobladas del Sur de Europa and
Centro de Estudios sobre Despoblación de Áreas Rurales, 2017.

Pisani, Donald J. *From the Family Farm to Agribusiness: The Irrigation Crusade in California
and the West, 1850–1931.* University of California Press, 1984.

Polo y Catalina, Juan. *Censo de la riqueza territorial e industrial de España en el año 1799,
formado de orden superior.* Madrid: Imprenta Real, 1803. https://hdl.handle.net/2027
/chi.096208450.

Pulido, Laura. "Geographies of Race and Ethnicity II: Environmental Racism, Racial Capitalism and State-Sanctioned Violence." *Progress in Human Geography* 41, no. 4 (August
2017): 524–33. https://doi.org/10.1177/0309132516646495.

Qiong Hu, Mingtao Xiang, Di Chen, Jie Zhou, Wenbin Wu, and Qian Song. "Global Cropland Intensification Surpassed Expansion Between 2000 and 2010: A Spatio-Temporal
Analysis Based on GlobeLand30." *Science of the Total Environment* 746 (December
2020): 141035. https://doi.org/10.1016/j.scitotenv.2020.141035.

Ramos, María E., Emilio Benítez, Pedro A. García, and Ana B. Robles. "Cover Crops Under Different Managements vs. Frequent Tillage in Almond Orchards in Semiarid Conditions: Effects on Soil Quality." *Applied Soil Ecology* 44, no. 1 (2010): 6–14. https://doi.org/10.1016/j.apsoil.2009.08.005.

Ramsey, Samuel D., Ronald Ochoa, Gary Bauchan, et al. "*Varroa destructor* Feeds Primarily on Honey Bee Fat Body Tissue and Not Hemolymph." *Proceedings of the National Academy of Sciences* 116, no. 5 (2019): 1792–801. https://doi.org/10.1073/pnas.1818371116.

Reisman, Emily. "The Great Almond Debate: A Subtle Double Movement in California Water." *Geoforum* 104 (August 2019): 137–46. https://doi.org/10.1016/j.geoforum.2019.04.021.

Reisman, Emily. "Plants, Pathogens, and the Politics of Care: *Xylella fastidiosa* and the Intra-Active Breakdown of Mallorca's Almond Ecology." *Cultural Anthropology* 36, no. 3 (2021). https://doi.org/10.14506/ca36.3.07.

Reisman, Emily. "Protecting Provenance, Abandoning Agriculture? Heritage Products, Industrial Ideals and the Uprooting of a Spanish Turrón." *Journal of Rural Studies* 89 (January 2022): 45–53. https://doi.org/10.1016/j.jrurstud.2021.11.003.

Reisman, Emily. "Superfood as Spatial Fix: The Ascent of the Almond." *Agriculture and Human Values* 37, no. 2 (2020): 337–51. https://doi.org/10.1007/s10460-019-09993-4.

Ricardo, David. *The Principles of Political Economy and Taxation*. Dover, 2004.

Ricart, Sandra, Anna Ribas, and David Pavón. "Modeling the Stakeholder Profile in Territorial Management: The Segarra-Garrigues Irrigation System, Spain." *Professional Geographer* 68, no. 3 (2016): 496–510. https://doi.org/10.1080/00330124.2015.1121834.

Riera, Francisco Javier, José Ferrán Lamich, and Juan Salom Calafell. *Cultivo del almendro*. 2nd ed. Barcelona: Aedos, 1970.

Riley, Elizabeth Margaret. "The History of the Almond Industry in California, 1850–1934." Unpublished master's thesis, University of California, Berkeley, 1948.

Rinkevich, Frank D. "Detection of Amitraz Resistance and Reduced Treatment Efficacy in the Varroa Mite, *Varroa destructor*, Within Commercial Beekeeping Operations." *PLOS ONE* 15, no. 1 (2020): e0227264. https://doi.org/10.1371/journal.pone.0227264.

Rivas Sanz, Juan Luis de las, and Miguel Fernández-Maroto. "Planning for Growth: Contradictions in the Framework of Economic and Urban Development from the 'Spanish Miracle' (1959–1973)." *Journal of Urban History* 49, no. 1 (2023): 41–59. https://doi.org/10.1177/0096144220983336.

Robinson, Cedric J. *Black Marxism: The Making of the Black Radical Tradition*. University of North Carolina Press, 2000.

Robledo, Ricardo. "La reforma agraria durante la Segunda República (1931–1939)." *Revista de estudios extremeños* 71 (2015): 19–48.

Romero, Adam. *Economic Poisoning: Industrial Waste and the Chemicalization of American Agriculture*. Critical Environments: Nature, Science, and Politics 8. University of California Press, 2022.

Romero Montero, Rafael. "La ayuda norteamericana a España." *Revista de estudios agrosociales* 30 (enero–marzo, 1960): 97–109.

Rominger, Robyn. "Navel Orangeworm: A Costly Pest in Almonds." *Farm Progress*, May 16, 2018. https://www.farmprogress.com/tree-nuts/navel-orangeworm-a-costly-pest-in-almonds.

Rosenstock, Todd, Daniel Liptzin, Johan Six, and Thomas P. Tomich. "Nitrogen Fertilizer Use in California: Assessing the Data, Trends and a Way Forward." *California Agriculture* 67, no. 1 (2013).

Rosiek, Jerry Lee, Jimmy Snyder, and Scott L. Pratt. "The New Materialisms and Indigenous Theories of Non-Human Agency: Making the Case for Respectful Anti-Colonial Engagement." *Qualitative Inquiry* 26, nos. 3–4 (2020): 331–46. https://doi.org/10.1177/1077800419830135.

Rubí, Ventura. *El almendro: Texto y fotografías.* Palma de Mallorca: Ministerio de Agricultura, Delegación de Baleares, 1980.

Rucker, Randal R., and Walter N. Thurman. *Combing the Landscape: An Economic History of Migratory Beekeeping in the United States.* North Carolina State University, 2019. https://ecg742.wordpress.ncsu.edu/files/2019/03/Combing-the-Landscape-complete-package.pdf.

Ruiz, Loreto Fernández. "Los nitratos y las aguas subterráneas en España." *Enseñanza de las ciencias de la tierra* 15, no. 3 (2007): 257–65.

Sacramento Daily Record-Union. "Suisun Valley: A Section of Remarkable Productiveness." January 28, 1888.

Sáez, Agustin, Marcelo A. Aizen, Sandra Medici, Matias Viel, Ethel Villalobos, and Pedro Negri. "Bees Increase Crop Yield in an Alleged Pollinator-Independent Almond Variety." *Scientific Reports* 10, no. 1 (2020): 3177. https://doi.org/10.1038/s41598-020-59995-0.

Sáez Pérez, Luis Antonio. "Análisis de la estrategia nacional frente a la despoblación en el reto demográfico en España." *Ager: Revista de estudios sobre despoblación y desarrollo rural / Journal of Depopulation and Rural Development Studies* 33 (October 31, 2021): 7–34. https://doi.org/10.4422/ager.2021.18.

Sala Roqueta, D. Ramon. "Sobre la polinización del almendro Desmayo." *Anales de la Escuela de Peritos Agrícolas y Superior de Agricultura y de los Servicios Técnicos de Agricultura* (1941): 43–56.

"Sample Cost to Produce Almonds: Glenn County 1963." College of Agriculture, University of California, 1963. https://coststudyfiles.ucdavis.edu/uploads/cs_public/e9/df/e9df45dc-7915-4649-9b36-6173a220211e/am-sv-63-almonds-1963-glenn_county.pdf.

Sánchez Martín, José Manuel, Juan Ignacio Rengifo Gallego, and Rocío Blas Morato. "Implantación de alojamientos en el medio rural y freno a la despoblación: Realidad o ficción. El caso de Extremadura (España)." *Revista de geografía Norte Grande* 76 (September 2020): 233–54. https://doi.org/10.4067/S0718-34022020000200233.

Sanden, B. L., A. E. Fulton, D. S. Munk, et al. "California's Effort to Improve Almond Orchard Crop Coefficients." Abstract, p. 7043. EGU General Assembly 2012, held April 22–27, 2012, in Vienna, Austria.

Sanden, Blake. "Fall Irrigation Management in a Drought Year for Almonds." *Kern Soil and Water.* University of California Cooperative Extension, September 2007.

Sanz, David, Santiago Castaño, Eduardo Cassiraga, et al. "Modeling Aquifer–River Interactions Under the Influence of Groundwater Abstraction in the Mancha Oriental System (SE Spain)." *Hydrogeology Journal* 19, no. 2 (2011): 475–87. https://doi.org/10.1007/s10040-010-0694-x.

Sapbamrer, Ratana, and Surat Hongsibsong. "Effects of Prenatal and Postnatal Exposure to Organophosphate Pesticides on Child Neurodevelopment in Different Age Groups:

A Systematic Review." *Environmental Science and Pollution Research* 26, no. 18 (June 2019): 18267–90. https://doi.org/10.1007/s11356-019-05126-w.

Saxton, Alexander. *The Indispensable Enemy: Labor and the Anti-Chinese Movement in California*. University of California Press, 1995.

Sayre, Nathan F. "The Genesis, History, and Limits of Carrying Capacity." *Annals of the Association of American Geographers* 98, no. 1 (2008): 120–34. https://doi.org/10.1080/00045600701734356.

Scales, Ivan R. "Green Capitalism." In *International Encyclopedia of Geography: People, the Earth, Environment and Technology*, edited by Douglas Richardson, Noel Castree, Michael F. Goodchild, Audrey Kobayashi, Weidong Liu, and Richard A. Marston. Wiley, 2017. https://doi.org/10.1002/9781118786352.wbieg0488.

Schatzki, Thomas F., and Martin S. Ong. "Dependence of Aflatoxin in Almonds on the Type and Amount of Insect Damage." *Journal of Agricultural and Food Chemistry* 49, no. 9 (September 2001): 4513–19. https://doi.org/10.1021/jf010585w.

Schnoor, Jerald L., ed. *Fate of Pesticides and Chemicals in the Environment*. Environmental Science and Technology. Wiley, 1992.

Scott, C. A., S. Vicuña, I. Blanco-Gutiérrez, F. Meza, and C. Varela-Ortega. "Irrigation Efficiency and Water-Policy Implications for River Basin Resilience." *Hydrology and Earth System Sciences* 18, no. 4 (2014): 1339–48. https://doi.org/10.5194/hess-18-1339-2014.

Sears, Louis, Joseph Caparelli, Clouse Lee, et al. "Jevons' Paradox and Efficient Irrigation Technology." *Sustainability* 10, no. 5 (2018): 1590. https://doi.org/10.3390/su10051590.

Segrelles Serrano, José Antonio. "La apicultura valenciana: Un aprovechamiento agrario tradicional." *Cuadernos de geografía* 45 (1989): 73–88.

Serrano, Ana, Ignacio Cazcarro, Miguel Martín-Retortillo, and Guillermo Rodríguez-López. "Europe's Orchard: The Role of Irrigation on the Spanish Agricultural Production." *Journal of Rural Studies* 110 (August 2024). https://doi.org/10.1016/j.jrurstud.2024.103376.

Shackel, Ken. "Precision Irrigation Management: What's Now and What's New (Part 1)—Water Production Function." Paper presented at the Almond Conference, Sacramento, CA, December 7, 2016.

Shubert, Adrian. *A Social History of Modern Spain*. A Social History of Europe. Routledge, 1996.

Siebert, John W. "Beekeeping, Pollination, and Externalities in California Agriculture." *American Journal of Agricultural Economics* 62, no. 2 (1980): 165–71. https://doi.org/10.2307/1239682.

Siebert, S., J. Burke, J. M. Faures, et al. "Groundwater Use for Irrigation—a Global Inventory." *Hydrology and Earth System Sciences* 14, no. 10 (2010): 1863–80. https://doi.org/10.5194/hess-14-1863-2010.

Simpson, Leanne. "Aboriginal Peoples and Knowledge: Decolonizing Our Process." *Canadian Journal of Native Studies* 1, no. 221 (2001): 137–48.

Smith, Kristine M., Elizabeth H. Loh, Melinda K. Rostal, Carlos M. Zambrana-Torrelio, Luciana Mendiola, and Peter Daszak. "Pathogens, Pests, and Economics: Drivers of Honey Bee Colony Declines and Losses." *EcoHealth* 10, no. 4 (2013): 434–45. https://doi.org/10.1007/s10393-013-0870-2.

Smith, Ryan, Rosemary Knight, and Scott Fendorf. "Overpumping Leads to California Groundwater Arsenic Threat." *Nature Communications* 9, no. 1 (December 2018): 2089. https://doi.org/10.1038/s41467-018-04475-3.

Socias i Company, Rafel, and Antonio Felipe. "Self-Compatibility in Almond: Transmission and Recent Advances." *Acta Horticulturae* 224 (1988): 307–17.

Socias i Company, Rafel, and Thomas M. Gradziel, eds. *Almonds: Botany, Production and Uses.* CABI, 2017.

Soluri, John. *Banana Cultures: Agriculture, Consumption, and Environmental Change in Honduras and the United States.* University of Texas Press, 2005.

Stafford, Raub Merrill. "Almond Survey of Sutter County." Bachelor's thesis, College of Agriculture, University of California, Berkeley, 1917. https://hdl.handle.net/2027/uc1.x86013.

Sterk, Geert, and Jetse J. Stoorvogel. "Desertification–Scientific Versus Political Realities." *Land* 9, no. 5 (May 18, 2020): 156. https://doi.org/10.3390/land9050156.

Stoll, Steven. *The Fruits of Natural Advantage: Making the Industrial Countryside in California.* University of California Press, 1998.

Strathern, Marilyn. *Partial Connections.* Rowman & Littlefield, 1991.

Suarez, Camille. "A Legal Confiscation: The 1851 Land Act and the Transformation of Californios into Colonized Colonizers." *Journal of the Civil War Era* 13, no. 1 (2023): 29–55. https://doi.org/10.1353/cwe.2023.0002.

Sundberg, Juanita. "Decolonizing Posthumanist Geographies." *Cultural Geographies* 21, no. 1 (2014): 33–47.

Sundberg, Juanita. "Feminist Political Ecology." In *The International Encyclopedia of Geography: People, the Earth, Environment and Technology*, ed. D. Richardson, N. Castree, M.F. Goodchild, A. Kobayashi, W. Liu, and R.A. Marston. Wiley, 2017. http://online library.wiley.com/doi/10.1002/9781118786352.wbieg0804/full.

Suryanarayanan, Sainath, and Daniel Lee Kleinman. *Vanishing Bees: Science, Politics, and Honeybee Health.* Nature, Society, and Culture. Rutgers University Press, 2017.

Swyngedouw, Erik. *Liquid Power: Water and Contested Modernities in Spain, 1898–2010.* Urban and Industrial Environments. MIT Press, 2015.

Swyngedouw, Erik. "Modernity and Hybridity: Nature, Regeneracionismo, and the Production of the Spanish Waterscape, 1890–1930." *Annals of the Association of American Geographers* 89, no. 3 (1999): 443–65. https://doi.org/10.1111/0004-5608.00157.

Swyngedouw, Erik. "'Not a Drop of Water . . .': State, Modernity and the Production of Nature in Spain, 1898–2010." *Environment and History* 20, no. 1 (February 2014): 67–92. https://doi.org/10.3197/096734014X13851121443445.

Taxin, Amy. "Crop-Rich California Region Will Fall Under State Monitoring to Preserve Groundwater Flow." Associated Press, April 16, 2024. https://apnews.com/article/cali fornia-groundwater-sustainability-tulare-lake-735900ab1de15e5eaa8000d263235125.

Taylor, R. H. *The Almond in California.* Bulletin 297 (August 1918). Agricultural Experiment Station, College of Agriculture, University of California, Berkeley. University of California Press, 1918.

Taylor, R. H., and G. L. Philp. *The Almond in California.* Circular 284. Agricultural Experiment Station, College of Agriculture, University of California, Berkeley. April 1925.

Taylor, Rebecca, and David Zilberman. "Diffusion of Drip Irrigation: The Case of California." *Applied Economic Perspectives and Policy* 39, no. 1 (2017): 16–40. https://doi .org/10.1093/aepp/ppw026.

Tello, Enric, Gabriel Jover, Ivan Murray, Onofre Fullana, and Ricard Soto. "From Feudal Colonization to Agrarian Capitalism in Mallorca: Peasant Endurance Under the Rise and Fall of Large Estates (1229–1900)." *Journal of Agrarian Change* 18, no. 3 (2018): 483–516. https://doi.org/10.1111/joac.12253.

Todd, Zoe. "An Indigenous Feminist's Take on the Ontological Turn: 'Ontology' Is Just Another Word for Colonialism." *Journal of Historical Sociology* 29, no. 1 (2016): 4–22. https://doi.org/10.1111/johs.12124.

Traynor, Joe. "A History of Almond Pollination in California." *Bee World* 94, no. 3 (2017): 69–79. https://doi.org/10.1080/0005772X.2017.1353273.

Tsing, Anna Lowenhaupt. *The Mushroom at the End of the World: On the Possibility of Life in Capitalist Ruins.* Princeton University Press, 2015.

Tsing, Anna Lowenhaupt. "Strathern Beyond the Human: Testimony of a Spore." *Theory, Culture & Society* 31, nos. 2–3 (2014): 221–41. https://doi.org/10.1177/0263276413509114.

Tucker, T. C. "The Future of the California Almond." Blue Diamond Brand, 1920. UC Davis Library Archives and Special Collections.

Tufts, Warren P., and Guy L. Philp. *Almond Pollination.* Bulletin. College of Agriculture, University of California, Berkeley, July 1922.

Tuin, Iris van der. "Diffraction as a Methodology for Feminist Onto-Epistemology: On Encountering Chantal Chawaf and Posthuman Interpellation." *Parallax* 20, no. 3 (2014): 231–44. https://doi.org/10.1080/13534645.2014.927631.

Turusov, Vladimir, Valery Rakitsky, and Lorenzo Tomatis. "Dichlorodiphenyltrichloroethane (DDT): Ubiquity, Persistence, and Risks." *Environmental Health Perspectives* 110, no. 2 (2002): 125–28. https://doi.org/10.1289/ehp.02110125.

Underhill, Vivian. "The Return of Pa'ashi: Colonial Unknowing and California's Tulare Lake." *Open Rivers: Rethinking Water, Place & Community* 24 (Fall 2023).

Underhill, Vivian, Sheeva Sabati, and Linnea Beckett. "Against Settler Sustainability: California's Groundwater as a Vertical Frontier." *Environment and Planning E: Nature and Space* 8, no. 1 (2022). https://doi.org/10.1177/25148486221110434.

United Nations Environment Programme. "Sustainable Consumption and Production Policies." Last updated June 18, 2021. https://www.unep.org/explore-topics/resource-efficiency/what-we-do/sustainable-consumption-and-production-policies.

University of California, Agriculture and Natural Resources. "Agriculture: Almond Pest Management Guidelines. Navel Orangeworm." UC IPM, June 2019. https://www2.ipm.ucanr.edu/agriculture/almond/Navel-Orangeworm.

University of California, College of Agriculture and Agricultural Experiment Station. *Report of the College of Agriculture and the Agricultural Experiment Station of the University of California from July 1, 1922 to June 30, 1923.* University of California Printing Office, 1923. https://hdl.handle.net/2027/uc1.b4334023.

US Climate Data. "Weather History Bakersfield—December 2019." Accessed 2023. https://www.usclimatedata.com/climate/bakersfield/california/united-states/usca0062.

US Congress House Committee on Appropriations. "Agriculture-Environmental and Consumer Protection Appropriations." In *Agricultural Appropriations for 1952: Hearings, Eighty-second Congress, first session, on H.R. 3973, making appropriations for the Department of Agriculture for the fiscal year ending June 30, 1952.* U.S. Government Printing Office, 1951.

US Department of Agriculture. *2020 California Almond Acreage Report.* National Agricultural Statistics Service, April 22, 2021.

US Department of Agriculture. "The Importance of Pollinators." https://www.usda.gov
/about-usda/general-information/initiatives-and-highlighted-programs/peoples-garden
/importance-pollinators.

US Department of Agriculture, Agricultural Marketing Service. "Almonds Grown in
California; Increased Assessment Rate." Proposed rule on July 18, 2016. Federal Register, 2016. https://www.federalregister.gov/documents/2016/07/18/2016-16814/almonds
-grown-in-california-increased-assessment-rate.

US Department of Agriculture, Agricultural Marketing Service. *Extracted Honey Grading
Manual*. Government Printing Office, 1985. https://www.ams.usda.gov/sites/default
/files/media/Extracted_Honey_Inspection_Instructions%5B1%5D.pdf.

US Department of Agriculture, Agricultural Research Service. *Beekeeping in the United
States*. Agricultural Handbook No. 335. USDA, 1967.

US Department of Agriculture, Division of Pomology. *Nut Culture in the United States,
Embracing Native and Introduced Species*. Government Printing Office, 1896.

US Department of Agriculture, National Agricultural Statistics Service, California Field
Office. *California Historic Commodity Data: Almonds 1909–2012*. February 2014.

US Department of Agriculture, National Agricultural Statistics Service. *Certified Organic
Survey 2021 Summary*. USDA NASS, December 2022.

US Department of Agriculture, National Agricultural Statistics Service. *Honey*. ISSN: 1949-
1492. March 2025.

US General Accounting Office. *The Role of Marketing Orders in Establishing and Maintaining Orderly Marketing Conditions*. Report to the Congress of the United States by the
Comptroller General, July 31, 1985. GAO/RCED-85-57.

Vallés y Vallés, Mariano. *El almendro: Su vegetación, zona de cultivo, terreno, variedades,
multiplicación, injerto, poda, abono, cosecha, enfermedades y enemigos*. 2nd ed. Barcelona: Libreria de Francisco Puig, 1932.

Van den Bosch, Robert. *The Pesticide Conspiracy*. University of California Press, 1989.

Van Leeuwen, Cynthia C. E., Erik L. H. Cammeraat, Joris de Vente, and Carolina Boix-
Fayos. "The Evolution of Soil Conservation Policies Targeting Land Abandonment and
Soil Erosion in Spain: A Review." *Land Use Policy* 83 (April 2019): 174–86. https://doi
.org/10.1016/j.landusepol.2019.01.018.

Vansell, G. H., and E. R. DeOng. *A Survey of Beekeeping in California and the Honeybee as
Pollinizer*. Circular. Agricultural Experiment Station, College of Agriculture, University
of California, Berkeley, October 1925.

Venot, Jean-Philippe. "From Obscurity to Prominence: How Drip Irrigation Conquered
the World." In *Drip Irrigation for Agriculture: Untold Stories of Efficiency, Innovation and
Development*, edited by Jean-Philippe Venot, Marcel Kuper, and Margreet Zwarteveen.
Earthscan Studies in Water Resource Management. Routledge, 2019.

Wade, William H. "Biology of the Navel Orangeworm, *Paramyelois transitella* (Walker), on
Almonds and Walnuts in Northern California." *Hilgardia* 31, no. 6 (1961): 129–71.

Wagner, David L., Eliza M. Grames, Matthew L. Forister, May R. Berenbaum, and David
Stopak. "Insect Decline in the Anthropocene: Death by a Thousand Cuts." *Proceedings
of the National Academy of Sciences* 118, no. 2 (2021). https://doi.org/10.1073/pnas.2023
989118.

Walker, Richard. *The Conquest of Bread: 150 Years of Agribusiness in California*. New Press,
2004.

Wallerstein, Immanuel. *The Modern World System 1: Capitalist Agriculture and the Origins of the European World-Economy in the Sixteenth Century*. University of California Press, 2011.

Wangxin Tang, Di Wang, Jiaqi Wang, et al. "Pyrethroid Pesticide Residues in the Global Environment: An Overview." *Chemosphere* 191 (January 2018): 990–1007. https://doi.org/10.1016/j.chemosphere.2017.10.115.

Warner, Keith. *Agroecology in Action: Extending Alternative Agriculture Through Social Networks*. Food, Health, and the Environment. MIT Press, 2007.

Watts, Vanessa. "Indigenous Place-Thought and Agency Amongst Humans and Non-Humans (First Woman and Sky Woman Go on a European World Tour!)." *Decolonization: Indigeneity, Education & Society* 2, no. 1 (2013): 20–34.

Weis, Anthony John. *The Global Food Economy: The Battle for the Future of Farming*. Zed Books, 2007.

Weiser, Matt. "Lucrative but Thirsty Almonds Come Under Fire amid Drought." *National Geographic*, April 21, 2015. https://www.nationalgeographic.com/history/article/140421-california-almonds-drought-central-valley-groundwater.

Wells, Miriam J. *Strawberry Fields: Politics, Class, and Work in California Agriculture*. Anthropology of Contemporary Issues. Cornell University Press, 1996.

Wey, Nancy. "A History of Chinese Americans in California." In *Five Views: An Ethnic Historic Site Survey for California*. California Department of Parks and Recreation, Office of Historic Preservation, 1988. https://www.nps.gov/parkhistory/online_books/5views/5views3.htm.

Whatmore, Sarah. *Hybrid Geographies: Natures, Cultures, Spaces*. SAGE, 2002.

Whyte, Kyle. "Settler Colonialism, Ecology, and Environmental Injustice." *Environment and Society* 9, no. 1 (2018): 125–44. https://doi.org/10.3167/ares.2018.090109.

Wickson, Edward. *California Illustrated: The Vacaville Early Fruit District*. California View, 1883.

Williams, Brian, and Jayson Maurice Porter. "Cotton, Whiteness, and Other Poisons." *Environmental Humanities* 14, no. 3 (2022): 499–521. https://doi.org/10.1215/22011919-9962827.

Williams, Peter, and Warwick E. Murray. "Behind the 'Miracle': Non-Traditional Agro-Exports and Water Stress in Marginalised Areas of Ica, Peru." *Bulletin of Latin American Research* 38, no. 5 (2019): 591–606. https://doi.org/10.1111/blar.12918.

Wood, Milo N. *Almond Culture in California*. Circular 103. California Agricultural Extension Service. College of Agriculture, University of California, Berkeley, 1937.

World Economic Forum and McKinsey & Company. *Innovation with a Purpose: The Role of Technological Innovation in Accelerating Food Systems Transformation*. January 2018, 42.

World Wildlife Fund. *Illegal Water Use in Spain: Causes, Effects and Solutions*. Report. WWF/Adena, May 2006.

Wynne, Brian. "Public Engagement as a Means of Restoring Public Trust in Science—Hitting the Notes, but Missing the Music?" *Community Genetics* 9, no. 3 (2006): 211–20. https://doi.org/10.1159/000092659.

Wynter, Sylvia. "Unsettling the Coloniality of Being/Power/Truth/Freedom: Towards the Human, After Man, Its Overrepresentation—An Argument." *CR: The New Centennial Review* 3, no. 3 (2003): 257–337. https://doi.org/10.1353/ncr.2004.0015.

Xingmei Liu, Yu Zhan, Yuzhou Luo, Minghua Zhang, Shu Geng, and Jianming Xu. "Almond Organophosphate and Pyrethroid Use in the San Joaquin Valley and Their Associated Environmental Risk." *Journal of Soils and Sediments* 12, no. 7 (2012): 1066–78. https://doi.org/10.1007/s11368-012-0519-8.

Yadav, Deepak, Rajasri Bhattacharyya, and Dibyajyoti Banerjee. "Acute Aluminum Phosphide Poisoning: The Menace of Phosphine Exposure." *Clinica Chimica Acta* 520 (September 2021): 34–42. https://doi.org/10.1016/j.cca.2021.05.026.

Yufang Jin, Bin Chen, Bruce D. Lampinen, and Patrick H. Brown. "Advancing Agricultural Production with Machine Learning Analytics: Yield Determinants for California's Almond Orchards." *Frontiers in Plant Science* 11 (March 13, 2020): 290. https://doi.org/10.3389/fpls.2020.00290.

Yu Zhan and Minghua Zhang. "Spatial and Temporal Patterns of Pesticide Use on California Almonds and Associated Risks to the Surrounding Environment." *Science of the Total Environment* 472 (February 2014): 517–29. https://doi.org/10.1016/j.scitotenv.2013.11.022.

Yuzhou Luo and Minghua Zhang. "Spatially Distributed Pesticide Exposure Assessment in the Central Valley, California, USA." *Environmental Pollution* 158, no. 5 (2010): 1629–37. https://doi.org/10.1016/j.envpol.2009.12.008.

Zierer, Clifford M. "Migratory Beekeepers of Southern California." *Geographical Review* 22, no. 2 (1932): 260. https://doi.org/10.2307/209177.

INDEX

Bold *page numbers refer to figures.*

activism, 17, 26, 38, 40

agribusiness, 3, 11, 13, 19, 30, 34, 71, 74, 108, 124

agrichemicals, 1, 3, 7, 9–15, 28, 56, 123, 132n65; and breeding, 38–40, 44–47; and plant-water relations, 51, 53, 56, 67; and pollinator dependence, 79, 83, 85–86, 94–95; and spatial precarity, 105–7, 119. *See also* insecticides; pesticides

agricultural extension services, 18, 26, 37, 39, 56, 60–63, 65, 80–81, 124

agronomy, 2–4, 9–13, 16, 19, 23–24, 27–29, 121–24; and breeding, 30, 39–40, 42, 46–47; and plant-water relations, 50–51, 58–64, 66–69, 71–75, 135n50; and pollinator dependence, 87–91, 94–95; and spatial precarity, 97–100, 102, 105, 107, 111, 115, 117, 119

air, 4, 13, 38–39, 43, 76, 85, 100–101, 105, 122; and plant-water relations, 58, 61, 63, 66

Alfonso XIII (king), 44

Algeria, 112

Almendrehesa, 116–18, 120

Almond Board of California, 3, 6–7, 11, 14–15, 23–26, 63, 66, 85, 94, 134n33; and spatial precarity, **99**, 102, 107, 109

almond industry, 4–8, 11, 14, 20, 23–25, 28–29, 133n81; and breeding, 34, 36, 40, 44–46; and plant-water relations, 50–51, 54, 57, 60, 73; and pollinator dependence, 82–87, 90, 94;

and spatial precarity, 98–99, 102, 105, 107–8, 113, 119. *See also* cultivation, almond

almond orchards, 1, 5, 7, 15, 25, **31**, 44, 46, 98, 105, 108; and plant-water relations, **49**, 53, 55, 61, 67, 69; and pollinator dependence, 76–78, 81–95

almond shells, 27, 33, 35–41, 44–47, 122, 131n23

Alqueva Dam, Portugal, 111

AlVelAl, 117

American Heart Association, 6

animals, 12–13, 38, 44, 79, 107, 115–17

anthropology, 21

aquifers, 9, 17–18, 28, 39, 102, 104, 107–8, 111, 119–23; and plant-water relations, 48, 52, 64

Argentina, 9

aridity, 1, 18, 25, 28, 89, 91, 97–102, 105–7, 112, 120–22; and plant-water relations, 48, 50, 52, 61, 66–67. *See also* dryness

Australia, 51, 60, 66

Ayerbe Castillo, Rafael, **31**, 40–44, 46–47

Barad, Karen, 21–22

biodiversity, 3, 79, 108, 114, 117

biophysics, 12, 32–33, 86, 92, 100, 121

birds, 37, 45, 107

Blue Diamond, 6, 26. *See also* California Almond Growers Exchange (CAGE)

bodies, 3, 13, 21, 39–40, 47, 82, 95, 101, 122

booms, almond, 4, 7, 9, 12, 14–15, 24, 26–29, 39, 44, 120–23; and plant-water relations, 51, 54,

163

Founded in 1893,
UNIVERSITY OF CALIFORNIA PRESS
publishes bold, progressive books and journals
on topics in the arts, humanities, social sciences,
and natural sciences—with a focus on social
justice issues—that inspire thought and action
among readers worldwide.

The UC PRESS FOUNDATION
raises funds to uphold the press's vital role
as an independent, nonprofit publisher, and
receives philanthropic support from a wide
range of individuals and institutions—and from
committed readers like you. To learn more, visit
ucpress.edu/supportus.